Routledge Introductions to Development
Series Editors
John Bale and David Drakakis-Smith

T0203730

Trade, Aid and Global Interdependence

The advent of multinational corporations and free trade zones has produced complex interrelationships between trade, aid and development. More than ever before trade has become a vital factor in the economic, political and social development of Third World nations.

Trade, Aid and Global Interdependence introduces trade as both concept and activity, placing aid within the context of trade in practice. The trend towards the globalization of trade, especially in the light of GATT and its emphasis on greater integration of economies throughout the world, means developing countries increasingly want both trade and aid. Aid alone is insufficient to bring about growth and economic development. Using a number of Third World case studies from Africa, Asia, Central and Latin America, the book analyses this competitive partnership of international trade and economic development, and discusses how various development strategies have been devised to respond to particular economic, social and environmental challenges.

Trade, Aid and Global Interdependence is a comprehensive introduction to the complex topic of international trade, and provides an important evaluation of global development – past, present and future.

George Cho is Senior Lecturer in Geographic Information Systems and Environmental Law at the University of Canberra, Australia.

In the same series

George Cho

Trade, Aid and Global Interdependence

Routledge
Taylor & Francis Group

LONDON AND NEW YORK

First published 1995
by Routledge
4 Park Square, Milton Park, Abingdon, Oxon OX14 4RN
605 Third Avenue, New York, NY 10017

*Routledge is an imprint of the Taylor & Francis Group, an informa
business*

© 1995 George Cho

Typeset in Times by J&L Composition Ltd, Filey, North Yorkshire

All rights reserved. No part of this book may be reprinted or
reproduced or utilized in any form or by any electronic,
mechanical, or other means, now known or hereafter
invented, including photocopying and recording, or in any
information or retrieval system, without permission in
writing from the publishers.

Notice:
Product or corporate names may be trademarks or registered
trademarks, and are used only for identification and explanation
without intent to infringe.

British Library Cataloguing in Publication Data
A catalogue record for this book is available from the British Library

Library of Congress Cataloguing in Publication Data
A catalogue record for this book has been requested

ISBN 13: 978-0-415-09159-6 (pbk)

To Marion and Carolyn

Contents

Plates

Figures

Tables

Acknowledgements

The author would like to thank the following for their kind permission to reproduce material for this book:

The Australian Department of Foreign Affairs and Trade, Canberra, for Plates 1.3, 3.1, 3.2, 3.3 and 3.4; the publisher and editor *Development Forum*, London, for Figure A.1 and Figure 3.2; the publisher, International Monetary Fund, New York, for Figure 2.1; M. Carr and Intermediate Technologies Development Corporation of America, New York, for Figure 3.3; the editor, *Sydney Morning Herald*, Sydney: John Fairfax Group, for Figure 4.2; the publisher and editors *The Australian World Atlas*, London: George Philip, for Figure 4.3; the publisher and editor *European Affairs*, London, for Figure 5.2; and the publisher and editors *Encyclopaedia of the World and Its People*, Sydney: Bay Books, for Figure 4.4 and 5.4.

I would also like to thank all the people who have assisted and encouraged me in the preparation of this book: without doubt my greatest debt is to Marion, my wife, especially for putting up with weekends and late nights and to Carolyn, my daughter, for giving me distractions and inspiration when I needed them most. Also to my students past and present who provided both a 'docile' and a critical audience which led to an improved presentation of this volume. *Merci beaucoup.*

Abbreviations

ADB	Asian Development Bank
AfDB	African Development Bank
AFTA	ASEAN Free Trade Area
AID	*Association Internationale de Développement (see IDA)*
AIDAB	Australian International Development Assistance Bureau (pre-April 1995)
AIJV	ASEAN Industrial Joint Venture
APEC	Asia-Pacific Economic Cooperation
ASEAN	Association of Southeast Asian Nations
AusAID	Australian Agency for International Development (post-April 1995)
CAP	Common Agricultural Policy
DAC	Development Assistance Committee
DFI	direct foreign investment
DIFF	Development Import Finance Facility
EC/EU	European Commission/European Union
EFTA	European Free Trade Association
EPZ	export processing zone
ESD	ecologically sustainable development
EU	European Union
FTZ	free trade zone
GATS	General Agreement on Trade in Services
GATT	General Agreement on Tariffs and Trade
GDP	gross domestic product

GNP	gross national product
GSP	generalized system of preferences
IBRD	International Bank for Reconstruction and Development
IDA	International Development Association (see AID)
IFAD	International Fund for Agricultural Development
IFC	International Finance Corporation
IMF	International Monetary Fund
ISI	import substituting industrialization
LDC	less developed country
MFC	most favoured customer
MFN	most favoured nation
MNC	multinational corporation
MTO	Multilateral Trade Organization
NAFTA	North American Free Trade Agreement
NGO	non-governmental organization
NIC	newly industrializing country
NIDL	new international division of labour
NIE	newly industrializing economy
NIEO	New International Economic Order
NTB	non-tariff barrier
ODA	official development assistance
OECD	Organization for Economic Cooperation and Development
OPEC	Organization of Petroleum Exporting Countries
R&D	research and development
SDR	Special Drawing Rights
UNCED	UN Conference on Environment and Development
UNCTAD	UN Conference on Trade and Development
UNCTC	UN Commission on Transnational Corporations
UNDP	UN Development Programme
UNHCR	UN High Commissioner for Refugees
VER	voluntary export restraint

1
Trade, aid and global interdependence

Introduction

Many countries seek to gain a foothold in global markets to sell goods in order to earn income. This income will help maintain a standard of living and a lifestyle which is compatible with the country's development and status. Developing countries also attempt to do the same by trading in global markets to produce the income that pays for development. But, for some, trade alone sometimes fails to produce enough to ensure growth and progress. Some developing countries therefore turn to international aid in one form or another – even the successful ones need to look for particular forms of aid – to help pay for very large projects. Together, trade and aid suggest the growing interdependence of countries – between the rich and poor, the developed and developing and the newly independent and sovereign states. This global interdependency is a significant feature of the 1990s given the breakdown of physical, cultural and ideological barriers since the Second World War. The removal of the Berlin Wall, the disintegration of the Soviet Union, the eradication of apartheid in South Africa are all signs of a change in thinking, in attitudes and in living in a world society. This book examines trade, aid and global interdependence and its implications for developing countries.

To trade is to exchange – goods for goods, money for goods and goods for money. But before trade takes place there is a need to assemble the goods from where they are produced and processed to points of sale where the exchange takes place. After the exchange, the goods are either further processed or consumed directly. Trade therefore may be thought of as the flow of goods and services over space. Also trade causes a division of labour as well as determining where goods are produced and where goods are consumed. Trade thus promotes the locational special-ization of an economic activity. This is because an activity will not take place in a local area if it does not enjoy some sort of an advantage. Economists call this 'comparative advantage', an advantage brought about by endowments of superior land, skilled labour, abundant capital resources and skilled managers. With trade, therefore, specialization, the division of labour and comparative advantage are important elements for earning income, for growth and development and for maintaining a certain lifestyle.

Thus, in order to understand the broader issues of trade, aid and global interdependence, it is necessary to examine each of these issues sepa-rately. By doing so, the interlinkages between the issues may be shown and the broader complexities of interdependence may be obtained. This chapter, first, discusses trade from the point of view of its main determinants, the spatial patterns that the flow of trade produces and the different theories of international trade which have been used to explain trade flow. Second, because trade takes place in a world with complex interrelationships, there is a need to consider past colonial linkages as well as the so-called North–South dependency relationships and that of unequal exchange. These relationships have emerged from the ideas involving the 'new international division of labour' (NIDL) and the 'New International Economic Order' (NIEO) – ideas arising from the need to restructure to redefine and reorganize international trade relations. Third, the context of regional integration, regional markets and trading blocs suggest the need to consider the role of the General Agreement on Tariffs and Trade (GATT) and the role of multinational corporations (MNCs) in global integration, in technology transfer and in their impact on the environment. An outline of the overall structure of this book concludes this introduction.

Trade is more than . . .

Trade is more than the mere exchange of goods and services. In the modern world, trade more often than not sows the seeds for growth and development. Here we refer to growth as additions to general well-being whereas development is the improvement in the general welfare and quality of life. Trade also provides the knowledge and experience that makes development possible. In turn, trade produces the capital goods that are indispensable for economic growth. Trade also brings together the know-how as the means and vehicle for the dissemination of technical knowledge and transmission of innovations. Along with the goods and services come the skills, managerial talents and entrepreneurship. Trade is also responsible for the transfer of capital from developed to developing countries, although invariably giving rise to some form of dependency relationships. Free trade in a liberal trading world can guarantee competition and avoid monopolies. If trade is truly free, there should be an optimal use of the world resources deriving from greater efficiency, internal and external economies of scale and thus smaller resource use per unit of output. Economies of scale include the reduction of unit production costs, the accumulation of reserves and operating on the principle of multiple and bulk purchases. Production is integrated both vertically and horizontally thereby enabling economies of scale to take place. Also production may be linked in marketing: backwards in buying the raw materials and forwards in selling the finished products.

For many countries international trade has been the principal link between nations. International trade has provided a powerful mechanism for the transmission of economic growth and the possibility of development. International trade is the principal means of extending markets beyond a nation's borders, thereby allowing greater specialization in production, enhanced effectiveness in use of scarce resources, the expansion of national income, the capacity to accumulate wealth and foster growth of the economy. Domestic welfare and economic growth can be unlimited where free trade is permitted between nations. Opportunities for more efficient production and greater output from each unit of resource may result from trade.

Trade activity is often facilitated by efficient modes of transport and is independent of political divisions. The sale of raw materials such as iron ore, or the import of finished products all depend on an efficient set of integrated transport systems. Such a demand for transport depends on two components – the demand to move a particular volume of goods and the distance over which it is carried. Except for bulk freight such as crude oil, coal and grain which use bulk carriers, internationally standard containers are now used to move goods quickly and with little loss

Plate 1.1 Trade and transport. Trade activity at a container port in Port Klang, Malaysia. The movement of goods in containers has many advantages over conventional methods including speed and efficiency and minimal losses through theft and breakages
Photo: G. Cho

through theft and breakage. Systems of transport therefore play an important role in fashioning the geography of the world by influencing the distribution of agriculture, the location of industries and the pattern of human settlements and work (see Plates 1.1–1.3).

International trade is thus seen here as the exchange of goods and services between distinct and separate political units. Such trade is also particularly amenable to control and taxation that yield revenues. Also by judicious regulation, trade directly influences the nature and extent of economic activity and indirectly the patterns of specialization within the nation and elsewhere. Curbs on international trade may result from the need to exclude prohibited goods such as drugs and other contraband. On the other hand, on economic grounds controls may be in the form of 'mercantilism' – the attitude that money is the only form of wealth, and the desire to either protect a nation's industries or to exert the power of the state in the marketplace.

Apart from government interference, a barrier to effective interna-

Plate 1.2 Containerization – a major development in the transport of goods permits quick loading and unloading. Containers are all of an internationally agreed size
Photo: G. Cho

tional trade is the geographical factor of distance. Two nations may be better off trading with an intervening country – the so-called intervening opportunity – if the friction of distance places too great a cost on the final product. The factors which influence and affect international trade thus include the physical and economic endowments and characteristics of trading nations, distance between countries, the political element, population and human resources and the stage of development of nations.

For a long time, the economist Ragnar Nurske's idea that trade can act as an 'engine of growth' for developing countries held sway. The idea was that the growth of developing countries was largely determined by external conditions. Economic growth was linked, on the one hand, by exports as well as imports of investment goods, and on the other, to the growth and prosperity of developed countries. Historically this explanation was plausible while many developed European nations depended on

Plate 1.3 Iron ore carrier leaving Hammersley Iron wharf at Dampier, Western Australia. The export of raw natural resources facilitates secondary and final processing at points of consumption. Supertankers can be loaded and unloaded quickly
Photo: Australian Department of Foreign Affairs and Trade

their colonies for raw materials for use in domestic industries. Economic expansion of developed countries also produced growth in developing countries. By the twentieth century, with independence and the emancipation of colonies, the engine of growth explanation became less tenable. Moreover, a fall in demand for raw materials and deteriorating terms of trade had made the idea even less convincing. A diversification of exports from developing countries coupled with the authority of the marketplace and price mechanism has undermined any support for the engine of growth idea. The contribution of agriculture to the gross domestic product (GDP) of growing Asian and Pacific countries has fallen dramatically, for example, in Thailand, where the contribution of agriculture fell from 32 per cent in 1965 to 15 per cent in 1989, while in Korea the corresponding figures are 38 per cent and 10 per cent and for Papua New Guinea are 42 per cent and 28 per cent over the same period.

The reciprocal relationship between trade and the location of economic activities explains the volume, direction and composition of trade between two areas. This also directly reflects the location of particular types of production and the markets for such goods. In this view, trade is seen primarily as a means of increasing economic welfare (that is, development) by facilitating specialization which in turn increases the efficiency of resource use. Trade is also a powerful mechanism for economic growth. In recent times, however, a worsening of the terms of trade for primary commodities has meant that trade has become unreliable as a source of growth. Moreover, international trade is often cited as one of the principal factors promoting the emergence of a hierarchically organized world system that is divided into a core, periphery and semi-periphery. To avoid such a world system, a shift from the mere palliative of import substituting industrialization (ISI) to one of integration with the global economy, including the absence of import protection and trade barriers of any description, is taking place. The import substitution approach ensured that domestic industries could grow, the nation would conserve scarce foreign capital within the country, decrease external dependencies and strengthen nationhood. But this approach meant a highly protected local market with resources allocated to protect infant industries – a theme discussed in Chapter 2.

International trade theories

Theories of international trade in general emphasize the salient elements of a complex system. The emphasis is on general rather than particular trade flows. The twin questions of what are the determinants of trade and what are the spatial patterns of trade suggest the need to examine various theories of international trade in order to obtain some answers.

The development of international trade theory has undergone four distinct phases. The first phase is the pre-classical or mercantilist following the Middle Ages till the 1750s or so. This was followed by the classical phase, lasting for about 150 years and coincident with the Industrial Revolution in Europe; then came the modern or contemporary trade theory and finally the neo-classical phase. An understanding of these four phases will permit a better explanation of trade policies adopted by different countries. (See Figure 1.1 for a synoptic view of the four phases.)

Pre-classical or mercantilist phase

Trade theory during the mercantilist period may be interpreted more as a set of policies of self-interest and economic nationalism rather than anything else. The features of such policies included favourable balance of trade with more exports than imports, the emphasis on foreign trade with manufactured goods rather than agricultural produce, and the use of controls and force to implement these policies.

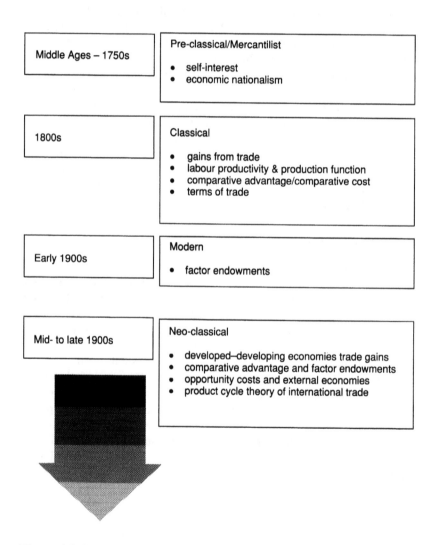

Figure 1.1 Summary of international trade theories

Classical phase

Economists such as Adam Smith and David Ricardo emphasized gains from trade. The premise was that free trade was beneficial to all trading partners and that the amount of goods to be traded depended on relative price levels. Trade resulted in increasingly specialized production. The price charged for any good depended on the units of labour used. A country will export those products for which labour productivity is high. Such productivity was dependent upon natural conditions in various countries so that labour in one country may be more productive in certain goods than in another. The economic concept to describe these differences is known as the production function which is assumed to vary between nations. Labour was completely mobile within a country but immobile between countries. Note that the cost of transport is not considered. The emphasis on one factor of production – labour – ignores the importance of other factors of production. The core of the theory of comparative costs is that trade occurs because comparative costs differ among countries. Comparative costs are defined here as the ratio of the production costs of the commodity distinguished in terms of the quantities of inputs.

David Ricardo's theory is based on comparative advantage or cost. That is, even though a country may have an absolute advantage over others in the production of every good, it would be in the interests of itself and other nations to specialize in producing and exporting those commodities in which it has a comparative cost advantage and importing those goods in which it has a comparative cost disadvantage. While the theory suggests reasons for the direction of trade flows it does not explain the actual amount of the final price. In international trade this is known as the terms of trade, that is, the ratio of the value of exports to imports relative to a base reference shown as a percentage figure.

Developing countries are particularly susceptible to poor terms of trade. This is because comparative costs are largely determined by the natural endowment of a country. Primary produce comes under this category. Also called Ricardo goods, the demand for and trade in primary commodities are dependent on global economic activity. Developing countries have particular difficulty in catching up with the developed world because the terms of trade are consistently working to their disadvantage. Moreover, technical change in the developed world has increased the number of substitutes available thus further reducing the demand for and price of raw materials originating from developing countries.

Today we observe a change in comparative advantage from purely geographical determinants to one of technical determinants. This change favours countries whose institutional structures and capital availability are adaptive to technical innovations. MNCs complicate the picture by controlling the export of primary commodities from developing countries. MNCs may remit income earned by such exports, thus reducing the capital available for internal industrial expansion within the developing country. An inequality in income distribution will witness the import of luxury goods to satisfy the demands of the wealthy groups, thus damaging the delicate balance of payments of the developing country. Finally, the reliance on primary products with low elasticity of demand, that is, any increase in price of the produce will result in the substitution of natural material for synthetic and industrially produced products, will affect the terms of trade.

Modern trade theory

Modern trade theory attempts to deal with the complexities of real-world international trade. This follows the abandonment of the labour theory of value in favour of including other factors of production such as land, capital and entrepreneurial skills. So differences in 'factor endowments' may determine the nature of trading relations between countries. Depending on the abundance or scarcity of factors of production, each country produced trading goods for which there was a natural advantage to do so. The starting premises for this theory are, first, countries differ in their factor endowments and, second, commodities differ in the combination of factors required, that is, factor intensities. Assuming that factor intensities of a given commodity are the same for different countries, this theory holds that each country will export those goods whose production is relatively intensive in the country's abundant factor and import those which are intensive in the factors it lacks.

Neo-classical trade theory

Neo-classical approaches predict trade will be most intensive between economies that differ widely and the greatest gains will be where conditions between the two countries are most dissimilar. The implication is that trade will be greatest and gains largest between developed and developing countries. This is because factor endowments and costs of production differ greatly between the two sets of countries. This also means that developed countries will export capital-intensive goods to developing countries who will export labour-intensive goods in return.

The Eli Heckscher–Bertil Ohlin model of international trade attempts to account for comparative advantage in terms of international differences in factor endowments: capital (including human capital), labour, resources, management and technology. A nation's comparative advantage will be determined by the relative abundance of its several factors of production. A country will export those goods which use a relatively large proportion of the factor that is relatively abundant, say labour or capital, and import commodities which use a relatively great deal of scarce factors (also known as Heckscher–Ohlin goods). This theory therefore shows a link between a nation's economic structure and its trade.

Further, the concept of opportunity cost is used where one commodity is substituted for another based on their relative costs, that is, substituting a cheaper good for a dearer one. The existence of external economies helps to explain trade between areas that do not have significant physical or cultural differences. External economies accrue when efficiencies gained by one country from trade are due to the expansion of world trade. The exchange of manufactured goods among countries of western Europe is such an example. Neo-classical theory thus concentrates on the determinants of the volume of trade, terms of trade, pattern of trade and on the connection between trade and the economic structure of trading countries.

However, the Heckscher–Ohlin theory has been criticized as being simplistic especially when applied to labour-scarce economies. For such economies, the theory predicts that there will be resource-intensive exports and labour-intensive imports. A cursory examination of Australia's trade pattern may suggest that this prediction is a reasonable one.

Moreover, the theory does not explain what is described as the product cycle theory of international trade. Labour is *the* most important factor of production. Here three stages of demand for any one commodity may be identified and labour plays an important role. In the first stage demand for the product is small and large-scale production may not be feasible. The inputs of skilled labour are large when compared to inputs of capital and unskilled labour. In stage two, as demand increases, more capital is used as an input together with managerial and engineering skills which are necessary for moving towards large-scale production. Then in the third stage of extensive demand, product standardization takes place where large amounts of capital and unskilled labour are combined with smaller amounts of skilled labour. Research and development may take place in one country while production takes place in another. These product cycle goods are highly sensitive to cost differences and tend to be produced where labour costs are still relatively cheap. (See also the detailed discussion on product cycle goods in Chapter 4, p. 110ff.)

To summarize, these various theories of international trade attempt both to explain and to predict the volumes and direction of trade between any two sets of countries. A geographical approach to the study of trade will examine not only the determinants of international trade but also the effects of trade on particular regions. In doing so, a better understanding of the contemporary geography of developing countries may be obtained. The analysis of the empirical evidence would provide an 'explanation' for the need for aid and the 'globalization of production' through trade.

Aid

Foreign aid is the international transfer of public resources and funds in the form of loans or grants either directly from one government to another (bilateral) or indirectly through an agency such as the World Bank. When foreign aid takes the form of the transfer of expert personnel, technicians, scientists, educators, economic advisers and consultants, then the term used is technical assistance. Where foreign aid is in the form of bilateral loans or grants which require the recipient country to use the funds to purchase goods and/or services from a donor country, the aid is said to be 'tied' to purchases from the donor country. Official development assistance (ODA) consists of net disbursements of loans or grants made at concessional financial terms by official agencies by members of the Development Assistance Committee of the Organization for Economic Cooperation and Development (OECD) and members of the Organization of Petroleum Exporting Countries (OPEC) with the objective of promoting economic development. In 1989, ODA on highly concessional terms made up 90 per cent of all grants and net lending from all official sources. Net disbursement of ODA from all sources for 1989 was US$41 billion compared to US$30 billion in 1983. It is the principal form of resource transfer to the poorest countries, accounting for nearly two-thirds of new resources flows to low income countries and four-fifths of flows to the poorest countries.

In the 1960s the catchword was 'trade not aid' whereas in the late 1980s the case was for 'trade through aid'. However, not all countries have been able to partake in either of these. Many developing countries are unable to trade because their productive capacities are undeveloped and therefore are unable to underwrite the costs of development. A lack of foreign exchange to purchase capital goods and equipment has resulted in the under-exploitation of potential capacity. Development projects initiated with foreign aid might have a significant impact on exploiting production capacities to their fullest extent.

The World Bank or, to give it its formal title, the International Bank for Reconstruction and Development (IBRD), aims to raise the standards of living in developing countries by redirecting financial resources from developed countries. World Bank loans generally have a grace period of five years, and are repayable over 20 years or less. One arm of the World Bank, the International Development Association (IDA), provides funds called credits (to distinguish these from World Bank loans) to governments. These credits have a 10-year grace period, mature in 50 years and carry no interest. An analysis of World Bank and IDA lending to countries in South East Asia reveals that the proportion of loans and credits is rather small relative to world-wide operations. World Bank loans to South East Asia amounted to US$21.2 billion, about 15 per cent of all loans, and US$2 billion in credits in 1987. World Bank policy reserves credit to help 'poorer' developing countries in order to conserve these countries balance of payments relative to the more 'expensive' World Bank loans (see Table 1.1).

The World Bank has stated that aiding well-conceived and well-monitored programmes makes a difference to the overall development effort. There can be no denying that the value of foreign aid has been significant, for instance, in the family planning programmes in India as

Table 1.1 Official development assistance

	World Bank / IBRD*		IDA
Form	grants	loans	credits
Grace period	nil	5 years	10 years
Repayment	nil	20 years	50 years
Interest rate	nil	low	none

Note: * IBRD – International Bank for Reconstruction and Development.

well as in other spheres of economic activity. Yet, external factors beyond the reach of both local policy-makers and donor countries may intrude with unfortunate consequences. The pervasive influence of lending agency policies is a much-vexed problem since it may lead to economic distress in receiving countries. If the objective is to make aid more effective, better domestic economic policies and direct foreign investments are equally important. Such effectiveness in turn depends on the policies of lenders and borrowers alike.

   While aid may have multiple objectives, especially when driven primarily by political considerations, donors need to ensure that the economic outcomes are achievable. As foreign aid can support and reinforce both good and bad domestic economic policy it is important recipient countries ensure that such policy is carefully developed. Human, financial and administrative infrastructure is required if foreign aid is to be used effectively. The volume of funding should be stable and the conditions of use of such funds set out clearly. This will help recipients make better use of the financial assistance. All these steps help make a difference in the quality and quantity of aid needed in particular countries.

Global interdependence

Geographers focus on the role of space and location in influencing the form and function of the international trading system. Such a spatial perspective divides the world into different camps – the developed and the developing worlds – in its attempts to explain the morphology of the international trade pattern. Yet, the explanations are unsatisfactory because the pressing questions of volume, terms of trade and patterns of trade remain unanswered. The more relevant aspects of theory which are important to geographers include an explanation of why countries engage in foreign trade, how trading partners are selected, and what commodities are traded in and in what quantities. (See Figure 1.2 for a perspective of the international trading system and global interdependence.)

Trade and dependency

Previously it was suggested that trade was an engine of growth for many developing countries. However, trade is also a mechanism for dependency. By growth is meant export expansion, while dependency implies that one group of countries controls indirectly the growth and expansion

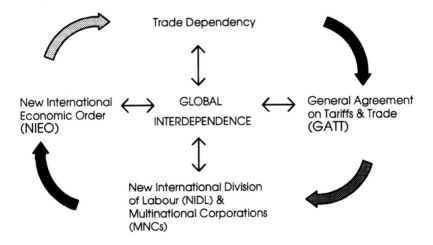

Figure 1.2 Spatial perspectives of the international trading system and global interdependence

of another group. A heavy dependence on developed countries as trade partners will mean that developing countries become sensitive to changes in the economic climate of their trading partners. The situation may also arise where a developing country will rely on a developed country's domestic and international economic policy in order to stimulate their own economic growth. Not only that, developing countries may even adopt the developed country's educational systems, their technology, economic and political systems, attitudes and consumption patterns in order to assume a developed status.

Some writers have explained that the deterioration in the terms of trade is the result of the system of unequal economic relationships between rich and poor countries. The resulting trade dependency is seen as an inevitable outcome of the process of international capitalism. This so-called neo-colonial model of underdevelopment postulates that under-development exists because of the exploitative economic, political and cultural policies of developed countries. This 'explanation' in the post-colonial period is characterized by inappropriate transfers of technology, unequal trading relationships and misdirected assistance programmes. Such conclusions may be over-simplified and may lack empirical merit. The effects of trade dependency, however, may be manifest in the nature of foreign trade, external payments, imported capital and technology and decision-making. More particularly, if a careful study is made of the economic structure of a trade dependent country, certain features and characteristics will be very evident.

First, the economy of a trade dependent nation will be small in size. Second, there will be a need for specialization in foreign trade precisely to exploit comparative advantage in the commodities it possesses. Third, there is a high ratio of international trade compared to the gross national product. Fourth, there will also be a high commodity concentration since it is a highly specialized producer. Finally, there is a high geographic concentration in the export destinations for historical rather than economic reasons. Such observations, however, should not be over-emphasized since there can be variations from country to country. Indeed, a different interpretation could also be put on these patterns. For instance, Marxist interpretation could argue that the growth of unequal economic relationships between the 'rich' and 'poor' countries (paralleling the conventional developed and developing model) is part of the process of international capitalism. Whatever interpretation one chooses to put to these processes one must always be cautious in using the concept of dependency. In the final analysis, the world is largely interdependent and must be taken as such.

The New International Economic Order (NIEO)

In the 1970s developing countries began to challenge the international economic order because they believed that the existing economic structures had been strongly stacked against their economic interests. These beliefs may be demonstrated by the growing corpus of evidence that shows how developing countries are being disadvantaged. First, there is an imbalance in the distribution of international monetary reserves. While the developing world constitutes 70 per cent of the world's population, they received less than 4 per cent of the international reserves of US$131 billion during the first half of the 1970s. The developed countries control the creation and distribution of such reserves and therefore also have the power to manipulate these financial reserves to their own benefit.

Second, developed countries control how and in what way the value-added products traded in the international marketplace are to be distributed. Developing nations usually receive only a small fraction of the final price obtained in the international marketplace. This may be explained by the fact that the control over processing, shipping and marketing of the primary products is achieved in the developed coun-

tries themselves. Developing countries may sometimes be in the paradoxical situation of repurchasing their 'own' primary products at highly
inflated prices for further processing because they had been unable to
process the materials initially.

Third, developed nations wishing to maintain their high standards of
living may introduce various forms of tariff and non-tariff protection of
their own inefficient domestic industries. Tariffs refer to taxes or duties
exacted against a particular group of merchandise entering or leaving a
country. Developing countries are therefore precluded from these markets because of the restrictive trade practices of these highly protected
international markets. Developing countries are unable to succeed
precisely because they have been excluded from participation in highly
lucrative markets.

Fourth, MNCs have 'engineered' their contracts, agreements and
concessions to their benefit at the expense of host countries. Such
countries are hardly in an equal bargaining position and more often
than not would welcome *any* kind of investment in order to simulate
the home economy. Moreover, MNCs, through the 'clever' use of tax
concessions, royalty payments, transfer pricing and capital allowances,
ensure that developing countries receive only a small fraction of the
profits of the venture after all necessary expenses have been disbursed.
Paradoxically, developing countries therefore get only a small fraction
of the benefits derived from the exploitation of their own natural
resources.

Fifth, developing countries have had little say in the decision-making
processes affecting the world economy. While representing a large
majority of the world's population, developing countries have only a
small representation on key economic institutions such as the World
Bank and International Monetary Fund (IMF). Even their numerical
majority in the UN carries little weight when international economic
decisions affecting developing countries are made.

Such inequities in international economic relations fuelled the desire
for developing countries to seek redress through the launch of the NIEO
in the 1970s. The UN General Assembly resolution of April 1974,
coming on the heels of the petroleum crisis, committed itself to:

> work urgently for the establishment of a new international economic
> order based on equity, sovereign equality, common interest and
> cooperation among all states, irrespective of their economic and
> social systems, which shall correct inequalities and redress existing

injustices, make it possible to eliminate the widening gap between the developed and the developing countries and ensure steadily accelerating economic and social development and peace and justice for present and future generations.

This declaration was adopted by all nations and spawned a series of 'programs of action'. What these sought to avoid was a new form of economic colonialism, given that many developing countries were only newly independent. The first UN Development Decade of the 1960s, while bringing development, also brought along unintended side-effects of rapid growth – population explosion in cities, depopulation of the rural areas and increasing socio-economic inequities within the developing country itself. The second UN Development Decade of the 1970s was implemented with more caution and with the petroleum crisis of the early 1970s, oil-rich developing countries found that they could command some attention in international economic relations. In a Programme for Action, NIEOs demanded immediate redress on four fronts. These included: aid and assistance, international trade, industrialization and technology, and social issues.

The main objective of aid and development assistance is to ensure that developed countries set aside a small proportion of their gross national product (GNP) for aid purposes. The goal for all developed countries is to increase assistance to an internationally agreed target of about 0.7 per cent of their GNP. In international trade the issues include redefining the terms of trade and access to markets. This is to be achieved through the use of a stable foreign currency with prices indexed for certain commodity prices to manufactured goods, to stabilize price fluctuations of primary commodities by establishing buffer stockpiles and preferential treatment of exports from developing countries. Industrialization issues include ISI, technology transfer, regulating the activities of MNCs and the improved use of natural resources. In social terms there were proposals to ensure an equitable distribution of income, the elimination of unemployment, provision of health, education and other cultural services for all people in developing countries. The NIEO is meant to ensure that developing countries can have some measure of economic independence. While there will inevitably be conflicts, it would appear that both developing and developed countries will gain immeasurably through cooperation. The world economic structure is now even more complex and intertwined so that each nation's growth and development are intimately bound up with those of other nations. The world is set to become even more interdependent in the new century.

Multinational corporations (MNCs) and the new international division of labour (NIDL)

An MNC is an international or transnational corporation which has its head offices in one country but branch offices in a wide range of both developed and developing countries. Examples include the Coca-Cola Corporation, Mitsubishi, Philips, IBM, British Petroleum and Exxon (see Figure 4.4 for an illustration of the various types of MNCs). These corporations are enormous in size and influence. Through various means of corporate mergers and company takeovers, MNCs have grown to a point where many now control large development budgets and their boardrooms negotiate with as much power as nations. In many cases they control or influence whole industrial processes from extraction of raw materials through to manufacturing, transport, finance and end use. According to a report by the UN Commission on Transnational Corporations (UNCTC) in 1991 the top 500 companies of the world now control about 70 per cent of world trade, 80 per cent of foreign investment and about 30 per cent of world GDP. MNCs control the way countries can and do develop while taking advantage of cheap labour. Together they control the economic and social performance of many countries, and also determine consumer tastes and patterns of consumption. But, apart from being the world's largest users of raw materials, they are also prime producers of goods that may be harmful to the environment.

Any discussion of the MNC must include its role in fostering inter-dependence. A fragmentation of production processes follows the establishment of MNC links so that there is now a global division of specialities, one country focusing on the extraction of raw materials, another on low-level processing and fabrication and a third on the assembly of components and final manufacture. Implicit in all these is the division of labour, the so-called new international division of labour (NIDL); although detractors suggest that the NIDL may in fact no longer be 'new'.

According to some there have been three recognizable periods for such division of labour and thus it may not justify the 'new' tag. The first NIDL occurred when colonies produced raw materials for export to metropolitan powers in exchange for manufactured and industrial goods. In the second NIDL, developing countries, particularly immediately after independence, launched themselves into import substituting industrialization (ISI). Such programmes were funded by both domestic and international capital. The third NIDL witnessed the fragmentation of the production process and the large flows of capital from within the core itself and from the core to the periphery. Such flows have largely been the result of the activities of the MNCs.

It is fascinating to note that all three types of NIDLs described above are currently in operation in various forms throughout the world. What impact each may have on the host nation will vary according to circumstance, although inescapably much capital investment is flowing into resource extraction projects, for example, Japanese investments in Indonesian timber resources. In contrast in Latin America, ISI has spawned local automotive industries which now go beyond assembling but have become the sole producers of vehicles that are no longer produced anywhere else in the world. For example, the Brazilian and Mexican Volkswagen are produced solely for the South American market.

The NIDL may be seen as the natural progression of trade, growth and development of different countries. The reduced capacity to generate profits within the metropolitan core has been the result of a rising wage bill, rising fuel costs and attendant costs to ensure occupational health and safety issues, together with pollution concerns. On the other hand, cheaper labour costs in the periphery, aided by eager governments aiming to achieve development at any cost, have acted as strong magnets. Also, because capital investments may have penetrated the countryside, a large army of displaced labour has found its way into urban areas, swelling the ranks of available labour and sustained by the informal sector. In addition, many developing countries actively court MNCs with special incentives usually by offering segregated economic zones such as free trade zones (FTZs) and economic processing zones (EPZs) with added 'tax holidays' for a period of time. MNCs take advantage of such lucrative inducements and often move on to other countries when these special incentives run out.

The nature of the product also has permitted the proliferation of assembly-type operations in developing countries. Modern technology has ensured that such operations are feasible so that, using sophisticated telecommunications such as satellite links, facsimile machines and computers, production may be coordinated and controlled from a metropolitan core. A parallel development is that the internationalization of financial services has resulted in accessibility of funds and their electronic transfer. Accumulations of capital are thus put to best use at all times with surpluses recycled many times over. The availability of capital to finance growth and development through official aid and assistance has resulted in a greater number of projects being started. Moreover, assistance through commercial sources using various devices of joint ventures and profit-sharing schemes has seen the flow of finances grow rapidly.

However, there are several criticisms of the NIDL as a concept. It is seen as holistic and all embracing and implies that the world can be divided up into two or three self-contained camps of the core, the semi-periphery and the periphery. The latter includes nations who are in receipt of MNC investment and who continue to engage in labour-intensive production. Moreover, the NIDL concept is geographically selective in that core capital is invested in only a very narrow range of countries for various reasons. The availability of resources, labour skills, existing infrastructure and type of government in power may explain the uneven patterns of investments. Within the developing countries themselves, investments may be highly selective, and there may not be a deep penetration of MNC capital. Local capital coupled with government investment may accompany foreign investments through deliberate policy and strategies, but the spatial impact of such investments is confined to urban centres of these developing countries.

The link between NIDL and global recession and expansion has not been established conclusively. Yet, there are some who see a direct link between the recession in core areas being transmitted to the periphery, the argument being that there is a time-lag for recession to occur in the periphery. While the apparent causation may seem to be clear, it is arguable if these are unrelated processes. There are obviously complex processes taking place which may be more than simply the result of the NIDL phenomenon. Explanations thus far are unconvincing.

The GATT and regional integration

Previous discussions of the NIDL and NIEO have underlined the importance of both economic and non-economic variables in this era of global interdependence. This observation also redefines such inter-dependence in which one nation's welfare may to some extent depend on the decisions and policies of another nation and vice versa. But the language used avoids the unproductive notions of superior–inferior,

dominant–dependent relationships. Rather the theme is one of cooperation and integration.

One of the dominant themes of the 1990s is the GATT. This international body was set up in 1947 to examine ways and means of reducing
tariffs on internationally traded goods and services. To achieve such
ends several conferences or 'rounds' of talks have been held to decide
on the levels of tariffs for primary commodities (see Chapter 5 and
Figure 5.3).

These conferences have met with moderate success. The Kennedy
Round (1964–7), named after President John F. Kennedy who had urged
the major industrial countries to undertake the negotiations, led to an
average reduction of about 35 per cent in tariffs on manufactured goods
and the introduction of rules against dumping. The Tokyo Round (1973–
9) yielded further reduction in tariffs on industrial goods and made it
harder for countries to use licences, customs rules and technical standards to keep imports out. The Uruguay Round (1986–93) is intended to
cut tariffs on a wide variety of goods, including, for the first time, a wide
array of agricultural products. It will also bring trade in textiles and
services under GATT jurisdiction. The agreement also calls for a new
more powerful Multilateral Trade Organization (MTO) to replace GATT.

The GATT multilateral trading system is founded on rules of equality
of opportunity in terms of access to markets and equality of treatment
inside national borders. Membership of GATT pledges a country to the
general expansion of trade on a multilateral and non-discriminatory
basis. The basic rules include non-discrimination among members
(except by members of free trade areas and customs unions) and that
protection should be by means of tariffs only, rather than quantitative
restrictions and similar devices. Taxes and regulation of domestic and
foreign producers should apply equally. These rules spell out key issues
such as multilateralism and free trade but avoid confronting regional
agreements. The GATT permits regional trade agreements as an exception to non-discrimination according to Article XXIV.

The GATT, however, is unable to find agreement between countries
concerning agricultural reform, fair trading and rules for protecting
intellectual property rights, equity and justice in the textile trade, lower
tariffs generally for all trade, dispute settlement procedures and trade
impacts of pollution on the environment. This list is a long one because
the issues are as contentious as they are lucrative financially.

The GATT may be seen as the engine of globalization. It is based on
the promise of a vigorous, profitable and peaceful commerce between

nations. Thus it is paradoxical that the GATT was in its element when the world was divided into two halves with the market economy of the West taking up one side and the command economies of the Soviet bloc taking up the other. Trade was promoted precisely to oppose the might of the other camp. However, in the climate of the 1990s, the distinctions have become blurred with the break-up of the Soviet Union, the economic and political integration of Europe through the European Community/Union (EC/EU) and the spawning of various regional trade blocs such as the North America Free Trade Agreement (NAFTA), Asia-Pacific Economic Cooperation (APEC) and ASEAN Free Trade Area (AFTA). The growth of these regional blocs could over time supplant and undermine global cohesion. Whether such recent developments are a good thing or not remains to be seen, but a failure of the GATT will mean greater regionalism.

Economic integration and regionalism appears to be an obvious solution to the problems of development for many smaller nations. By combining their markets and through economic cooperation it should be possible to improve their economic position. Theoretically, the logic of integration appears sound provided such nations are willing to accept that there will be the need to supplant national goals by supranational ideals. However, where there is a geographical spread over great distances, the dispersed pattern may create difficulties of communication

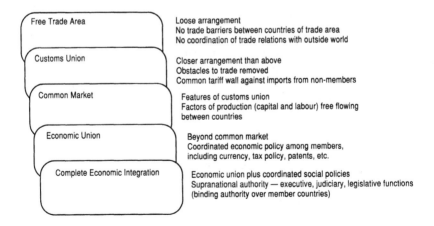

Figure 1.3 Models of economic integration

and integration. For some countries, this 'new colonialism' may be too high a price to pay for possible economic gains. The sharing of such benefits among integrated nations can also be difficult particularly where the contributions of each have been unequal.

There are several types of integration models including free trade areas, customs unions, common markets, economic unions and complete economic and political unions (see Figure 1.3). The logic is to bring about some measure of internal and external economies of scale. But as already pointed out, policies of trade integration often conflict with those of national independence with the possibility of a total industrial domination by one member creating new patterns of spatial inequality. In an Asian context, however, it may be argued that because decisions of member nations are made on consensus, often a long, slow and ponderous process, there will be little or no danger of any one country dominating decisions and hence determining new spatial patterns of industrial activity. But, this does not preclude regional inequalities within countries themselves.

The purposes of regional integration are to promote more trade among members through economic cooperation, to generate external trade and to raise the standards of living in each country. However, higher prices may be paid for imports from regional partners because production methods may be less efficient and by comparison the commodity is relatively more expensive than elsewhere in the world. Also the adoption of new technology will be retarded because of less competitive conditions, and the loss of potential gains from trade with the rest of the world arising from the regional union. On the positive side, regional groupings benefit from the elimination of barriers to trade by partner countries, thereby widening the market for primary and secondary industries. Trade integration may also increase the bargaining powers of participants in difficult trade negotiations. An improved flow of information about participating markets and competing markets reduces the risks and uncertainty in demand. As a unit the economic grouping as a whole permits the regional protection of 'infant industries' for which perhaps the national market itself may be either too small for efficient operation or too small to generate natural economies of scale.

Organization of this book

In Chapter 2 we will focus attention on the importance of international trade to the economies of developing countries. Following an introductory section on the relationship between trade and development, there is a discussion on the spatial structure and distribution of world trade. A sketch of broad trade patterns will permit a deeper understanding of the direction, composition and terms of trade among developed and developing countries. Such an understanding will help provide explanations of the various trade policies pursued by developing countries in the context of overall development strategies (ISI, export promotion, regional economic integration) and the problems associated with global trade reform such as GATT and the UN Conference on Trade and Development (UNCTAD).

As trade is sometimes unable to provide all the impetus required for the development effort, Chapter 3 examines the role of bilateral and multilateral aid to trace the flow of capital into developing countries. Such an examination will provide a profile of the volume of transfers, the donors and recipients and the achievements of such efforts. There will also be a discussion of the debt crisis and 'deficits' of many developing countries in order to appreciate how some countries have 'traded' themselves out of their fiscal difficulties.

In Chapter 4, we examine a particular mechanism which has been used by developing countries to help solve their balance of payments difficulties. Countertrade is a method of paying for goods and services in kind. This modern-day barter or exchange of goods and offset deals has increasingly been used by many debt-ridden countries to overcome their current difficulties. The problems and opportunities provided by such tactics are discussed. Also there is an evaluation of the process of transfer of technology in the light of attempts to reduce 'dependency' on developed countries.

Chapter 5 concentrates on the change from mere international trade to the internationalization of production with the emergence of NIDL. The prospects of aid and the NIEO are reviewed given current trends towards regional trading blocs and the prospects of the Uruguay Round of the GATT. This chapter draws conclusions from the previous chapters to produce a prognosis for the future.

Key ideas

1 Trade is more than the exchange of goods and services.
2 The volume, direction and composition of trade between two areas is the result of the geographical location of economic activity.

3 International trade theories attempt to explain the determinants of trade and the spatial patterns of trade.
4 While trade is the 'engine of growth' for many developing countries, trade is also a mechanism for 'dependency' in which one group of countries determine and condition the growth and expansion of another group.
5 Developing countries are challenging the international economic order as they believe that existing economic structures have been stacked against their economic interests.
6 Multinational corporations (MNCs) are fostering a new international division of labour (NIDL) by enlarging the geographical spread of production and consumption of goods and services.
7 Agreements such as the GATT are fostering freer trade and global interdependence.
8 The transfer of public resources and funds is helping accelerate development in the Third World.

2
Trade and development

Introduction

Any critical assessment of what is meant by 'development' should
include social and economic dimensions. In this respect development
means the qualitative improvement to welfare and standards of living.
Growth, on the other hand, is simply a quantitative improvement to
productivity and production of goods and services. Such an assessment
is in line with the approaches that have been taken by international
bodies like the World Bank and the UN Development Programme
(UNDP). Moreover, any serious study of development should actively
scrutinize the double-edged issues of distribution and growth, and be
concerned with the ecological as well as material aspects. The processes
of development are equally important so that the main beneficiaries of
development, the people themselves, can and do participate in determining their own 'development'. But development does not mean forsaking
one's own cultural and historical identities – indeed these aspects need
to be re-established and maintained since, as the Brandt Report (1980)
has emphasized, 'cultural identity gives people dignity'. To be developed any country must constantly strive for a high-quality economy
based on a coherent set of objectives and policies that produce social
justice and efficiency.

In the first part of the chapter we examine the nexus between
international trade and economic development, since trade contributes
to economic growth. Then in the second part the discussion switches to

an examination of the direction, composition, terms of trade and growth of trade for various regions of the world. The analysis provides a broad pattern of the globalization of economies and the complexities that these produce. By implication, it may be that economies are becoming more interdependent not only at a political level but also at the level of the multinational corporation (MNC). The third part of the chapter analyses the various development strategies and trade policies employed by Third World countries to enhance their economic development, and measures used to counter the negative aspects of interdependency.

International trade and economic development

A study of trade functions and international trade systems begins by asking why various countries perform different trade functions and how current trade patterns have emerged. The answers to such questions are usually based on economic notions of comparative advantage and differences between countries in the scale of their economies. However, to these may be added other non-economic notions such as differences in tastes, in factor endowments and in technologies. The geographer is driven by the need to understand precisely how spatial systems, as portrayed by trade patterns, operate. Such understanding is conditioned by a number of factors including the use of resources, location of such resources, the patterning and organization of the spatial economy, sometimes known as structure, and, as a result of all these activities, the growth and decline of particular regions.

While it is well known that trade flows depend on the existence of spatial and sectoral specializations in any economy, for example, the agricultural sector relies to a great extent on the industrial sector of the economy, there is still the need to put such notions into a conceptual framework. The use of spatial models that attempt to explain these relationships may be a useful beginning. A spatial model of a trading system should do the following: first, such a model should provide some explanation of the spatial distribution of production and consumption. Second, the interrelationships and spatial interactions manifested as trade and commerce should be simplified and portrayed in cartographic form. Finally, such a model should at least describe and explain regional differences in growth and development. These subscribe to the notion that the flow of goods and services in international trade indeed has some underlying logic. Such flows are the result of decisions made by individuals, households and firms who collectively determine the volume, direction and composition of trade.

Ron Johnston (1967) proposed a simple framework to discuss international trade patterns as a spatial system having three interacting components in which the system operates. First, there is a set of fixed elements which is characterized by certain attributes. These elements and attributes can be thought of as countries and economic production respectively. Then there is a set of relationships or interactions between the fixed elements. These linkages are established through formal links such as laws, contracts and understandings and through established structures such as physical transport and delivery, payment mechanisms and the resolution of disputes. The third set of relationships are those which connect the fixed elements with their external environment and other elements of the larger system. This framework gives structure to the system and also includes the notion of a causal system in which one element may cause another to respond in a particular way.

There have been many attempts to put these simple ideas into a workable scheme but no easy solution has been found. A consignment of goods threads a maze of great complexity. While it may be easy to think of international trade as an exchange between countries and economies, this simple view is inaccurate. Trade data of market economies represent the aggregation of countless individual transactions of exports and imports. In the command economies of socialist countries, state trading and government monopolies determine the volume of imports and exports. In between these extremes of economic organization are intermixed a myriad of private and public transactions.

The present task therefore is to map out international trade to detect geographical patterns. Previous discussions have suggested that the level of economic development may be a major determinant of the pattern of international trade with the hint that the geographical distance between potential trade partners may be a conditioning variable. Also it seems that developed economies are far more important for the developing world than the other way round. However, with the globalization of national economies such a notion may no longer be wholly true. A lack of trade between developing economies, the predetermined composition of goods exported and the physical transport network between two countries may increase the relative bargaining position of developing countries. Thus, it is imperative that the geographical pattern of international trade be described simply. One suggestion is to examine the geographical distribution of exports and imports, and then the commodity distribution of exports and imports – the basic 'who' and 'what' questions. Such a description is prompted by the stability in the patterns of international trade flows since the Second World War, despite the massive growth in the volume of world trade of about 25 times, and the dramatic change in trade patterns of individual commodities, such as petroleum and industrial raw materials, to counter declines in the movement of traditional energy goods. The changes in transport technology and the downturn in the trade for food items because of self-sufficiency, substitution of products and growth of manufacturing industries, may be contrasted with the growth of high-value, low-density commodities associated with the electronics and information-based industries.

The commodity composition and the direction of flows to and from most countries, while changing, may point up differences in levels of economic growth. Rather than deal with individual countries and in order to properly handle the description of global trade patterns, there is a need to group them into various categories. In preparation for the next section, the present task therefore is to examine how some of these categories have been derived in the context of development and growth of particular economies.

Development is taken here to refer to the process of improving the quality of all human lives. This includes raising the levels of living, the creation of conditions conducive to the growth of people's self-esteem and the freedom of choice by enlarging the range of their choices. Growth, on the other hand, is the process by which the productive capacity of the economy is increased over time to bring about rising levels of national income. Rapid economic growth is thought to be a major precondition 'determining' levels of living. Since the Second World War, throughout the world there has been rapid economic advance, with some developing countries making progress rapidly and others at a slower pace. Development is a multidimensional process depending on a complex interaction among institutions, policies and global economic climate. In setting out the priorities for action, the World Bank (1991) listed the responsibilities of industrialized countries and finance agencies. Foremost was the need to defend and extend the liberal order of international trade established after 1945, ease the flow of capital across borders, pursue domestic economic policies that promote global savings and non-inflationary growth, support the transfer of technology, protect the environment and conserve energy. In addition, the priorities for developing economies should include an investment in people, education, health and population control; help for domestic markets by fostering competition and investing in infrastructure; the liberalizing of trade and foreign investment; and the avoidance of excessive fiscal deficits and high inflation. History has shown that industrial countries have grown prosperous through trade. Ergo, there should be no effort spared to ensure that developing countries follow this same path to progress. For their own part, developing countries can play a key role by pressing for free trade and the continued reform of their own trade systems.

While by no means fully understood, one process is said to be driving economic development. Growing productivity has been identified as the engine of development. What drives this engine is the technological progress of the past decades as well as education, institutions and policies for openness in developing and industrial countries. There is strong evidence that links productivity to technological investments and to the extent to which markets are distorted by such factors as foreign exchange, price systems and restrictions on trade. Equally, where there is no growth in productivity there will be a decline in gross domestic product (GDP). The seven-year expansion in the world economy almost came to a halt in 1990. This led to tightened monetary policies and while production was at near capacity levels there was rising inflation. The Gulf Crisis of 1990 increased uncertainty as did the financial requirements of German reunification and the war-related reconstruction in the Middle East. For developing countries, the real GDP growth declined from 4.3 per cent in 1988 to 2.2 per cent in 1990. Such a decline has been traced to continuing macro-economic instability, domestic policy weakness, falling non-oil commodity prices, high international (non-US dollar) interest rates and a slower growth in world trade. The final point is most significant for present purposes because of its close relationship to both growth and development. According to the World Bank (1991) development is now seen as the most important challenge facing the human race. Despite the vast opportunities that have been created in the modern technological world, more than 1 billion people, one-fifth of the world's population, live on less than 1 US dollar a day – a standard of living that western Europe and the United States attained some 200 years ago.

There is now a greater reliance placed on markets to give the impetus to development. Countries which have achieved rapid development include those that have invested heavily in human capital through education and physical capital. Also markets, competition and trade have been given leading roles in order to achieve high productivity from investments (see Case study A).

To discuss the trade and market for goods and services meaningfully, it will be necessary to place countries into both country groups and analytical groups. This will allow the portrayal of broad patterns of relationships in international trade. The scheme follows the World Bank method of using country groups by classifying economies according to their gross national product (GNP). Every economy is classified as low-income, middle-income (subdivided into lower-, middle- and upper-middle) or high-income. In addition to income, other ways of classification are by exports, levels of external debt and regions.

Following the World Bank (1992):

- low-income economies are those with a GNP per capita of US$610 or less in 1990;
- middle-income economies are those with a GNP per capita of more than US$610 but less than US$7,620 in 1990. A further division at GNP per capita of US$2,465 in 1990 is made between lower-middle-income and upper-middle-income economies;
- high-income economies are those with a GNP per capita of US$7,620 or more in 1990.

Low- and middle-income economies are sometimes referred to as developing economies, which, although a convenient term, does not imply that all economies in the group are experiencing similar development. For analytical purposes, the World Bank uses an additional classification based predominantly on exports or external debt in addition to geographic country groups.

- Fuel exporters are countries for which exports of petroleum or gas, including re-exports, account for at least 50 per cent of exports of goods and services. Examples from the Gulf States include the Islamic Republic of Iran, Iraq, Libya, Oman, Saudi Arabia and United Arab Emirates.
- Severely indebted middle-income countries are fifteen countries that are deemed to have encountered severe debt-servicing difficulties (see Figure 3.1, p. 57).
- OECD members, a subgroup of high-income economies, comprises members of the OECD except Greece, Portugal and Turkey, which are included in the middle-income economies.

Case study A

GNP and world trade

Gross national product (GNP) and patterns of export trade reveal much about any economy and its relationships with other countries. It seems that countries with the largest GNPs are those that dominate international trade and have done so for the past 100 years or so. In recent times, however, new players to international trade have tended to dominate. A very distinct pattern of international trade is that about four-fifths of the trade flows between developed capitalist countries. The command economies of the socialist world and the developing countries are struggling to keep their share of an increasingly global and interdependent world economy.

Since the distribution of resources is uneven across the globe, the patterns of production vary from region to region as do the costs of such production. This produces inequities because of differences in comparative advantage. The pattern is thus one of specialization of production and hence of international trade. It has been argued that this is also the result of historical relationships between the developed and developing countries such as those based on colonialism. In recent times these have been based more on the activities of multinational corporations (MNCs), on direct foreign investments (DFI), import substituting industrialization (ISI), and export-oriented policies that are linked to the granting of most favoured nation status and trade concessions. The pattern in the early nineteenth century was the extraction of resources from less developed countries and their use and accumulation in the developed world. This free trade philosophy reflected the idea that less developed countries supplied the raw materials and resources while the developed world manufactured industrial goods. However, all these are now changing since resource-rich developing countries are able to command far better trade agreements with developed countries.

Yet, the continuing growth of international trade – by about 25 times since the Second World War – has done little to bring about a more equal distribution of wealth among nations of the world.

Case Study A (*continued*)

Such an unequal distribution of wealth is also reflected within many developing countries. The GNP provides a quantitative measure of relative wealth of a nation. The GNP per capita tells us about the relative wealth of individuals within a country. But, in reality, these two statistics conceal how a country's wealth is shared.

Figure A.1 Unjust terms of trade may prevent poor nations from developing
Source: *Development Forum*, vol. 3, no. 9: 7

Global trade

Developed countries, with only one-fifth of the global population, account for four-fifths of world output, more than four-fifths of world trade and almost all exports of capital and technology (World Bank 1991: 149). Any prospects for growth and development in developing countries will be strongly influenced by the performance of industrial countries. Developing countries are now turning outwards for trade, technical assistance and capital. In the 1950s and 1960s the US was acknowledged as the world's economic and political power. However, by the 1970s and 1980s this preponderant influence of the US had been lost and was now shared with Japan and western Europe. At the same time, north-east Asian countries have gained an increasing share of the world production of goods and services, rising from 8 per cent to 20 per cent while that of North America fell from 46 per cent to 32 per cent and western Europe increased from 29 per cent to 33 per cent (Garnaut 1989: 38). The rise of the 'Asian Dragons', which was based on the export of manufactured goods, has been especially dramatic. In the late 1980s, the Four Dragons – Singapore, Hong Kong, Taiwan and South Korea – produced 77 per cent of total manufacturing exports from developing countries. Such changes in international trade patterns therefore require detailed statistical description so as to identify the major players (see Figure 2.1).

A statistical description of global trade patterns may begin by examining the growth of (merchandise) trade for the decade of the 1980s (1980–90) (see Table 2.1). In 1990 world exports were valued at US$3.187 billion compared with imports of US$3.355 billion. The

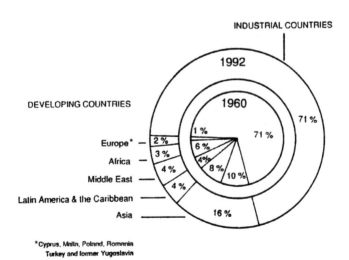

Figure 2.1 Relative percentage shares of world exports and imports in 1960 and 1992 for industrial and developing countries
Source: International Monetary Fund, *Direction of Trade Statistics Yearbooks* (various years)

Table 2.1 Growth of merchandise trade, 1980–90

	Exports 1990 US$ (m.)	Imports 1990 US$ (m.)	Average annual growth rate (%) exports 1980–90	Average annual growth rate (%) imports 1980–90	Terms of trade 1990 (1987 = 100)
Low-income economies	141,176	144,431	5.1	2.8	100
(China and India)	(80,059)	(77,037)	(9.8)	(8.0)	(103)
Middle-income economies	491,128	485,897	3.8	0.9	102
Lower-middle-income	184,340	195,680	7.2	2.1	99
Upper-middle-income	306,789	290,217	1.9	0.1	105
Sub-Saharan Africa	34,056	32,377	0.2	−4.3	100
East Asia and Pacific	217,030	224,021	9.8	8.0	103
South Asia	27,699	38,217	6.8	4.1	95
Europe	94,082	126,493	–	–	103
Middle East and North Africa	112,644	89,842	−1.1	−4.7	96
Latin America and Caribbean	123,181	101,119	3.0	−2.1	110
Severely indebted	135,856	99,721	3.4	−2.1	101
High-income economies	2,555,661	2,725,419	4.3	5.3	100
(OECD members)	(2,379,089)	(2,501,753)	(4.1)	(5.2)	(100)
World	3,187,955	3,355,746	4.3	4.5	100

Source: World Bank (1992) *World Development Report 1992*, New York: Oxford University Press, Table 14, pp. 244–5

dominance of high-income economies is particularly striking, accounting for nearly four-fifths of all exports and imports in 1990. The share of middle-income economies in global trade was just under 14 per cent, with low-income economies accounting for about 4 per cent. OECD members accounted for three-quarters of all trade in 1990. Between 1980 and 1990 the Middle East and North Africa recorded a negative growth rate in exports. This same group of countries also experienced negative growth rates in imports together with sub-Saharan Africa, Latin America and the Caribbean and the severely indebted economies.

Growth of trade

The average annual growth rate in percentage terms for the period 1980 and 1990 for exports and imports was about 4.3 and 4.5 per cent respectively. The growth in exports was particularly high for countries in East Asia and the Pacific (9.8 per cent) and South Asia (6.8 per cent), whereas in countries in Latin America and the Caribbean (3 per cent) and in the severely indebted countries (3.4 per cent) the growth has been unspectacular. Indeed, sub-Saharan Africa recorded an average annual growth of just 0.2 per cent and the Middle East and North Africa recorded a negative growth of −1.1 per cent. In terms of imports, China and India show the highest growth rates of 8 per cent with East Asian and Pacific countries recording 8 per cent. There has been negative growth of imports for countries in the Middle East and North Africa (−4.7 per cent), the sub-Saharan African group (−4.3 per cent), Latin America and Caribbean (−2.1 per cent) and the severely indebted (−2.1 per cent).

Terms of trade

The terms of trade refers to the fact that if the price of a country's imports increases while that of its exports remain the same, then for a given amount of exports, that country can purchase only a diminished amount of imports. In general, the terms of trade has deteriorated for all countries except those comprising South Asia, the Middle East and North Africa where the index is 95 and 96 respectively. In fact Latin America and the Caribbean and severely indebted countries recorded very high changes in the terms of trade index: 110 and 101 respectively. This means that compared with the base year (1987) the value of the export trade has deteriorated relative to the value of the import trade.

Composition of trade

Table 2.2 gives the structure of merchandise imports in 1990 and shows that machinery and transport equipment, together with 'other' manufacturing feature heavily for all economies. These imports account for between 20 and 40 per cent and 34 and 41 per cent respectively of all imports. This is not surprising considering a majority of such imports may be classified as 'capital goods' and consumer-oriented products. Food imports world-wide are about 9 per cent with fuels (11 per cent) and other primary commodities (8 per cent) making up the rest of the import trade. There are no great variations in the mix of imports for all economies under consideration, with only the sub-Saharan African countries importing less fuels and other primary commodities.

In terms of the structure of merchandise exports for 1990, Table 2.3 shows the predominance of middle- and high-income economies in the export of machinery and transport equipment and in other manufactured goods. In fact, the trade in these goods dominate the global pattern

Table 2.2 Structure of merchandise imports, 1990 (percentage share of merchandise imports – weighted averages)

	Food	Fuels	Other primary commodities	Machinery and transport equipment	Other manufactures
Low-income economies	12	9	8	33	38
(China and India)	(8)	(7)	(10)	(34)	(41)
Middle-income economies	11	12	8	34	35
Lower-middle-income	11	10	8	34	37
Upper-middle-income	10	13	9	33	34
Sub-Saharan Africa	16	4	3	40	37
East Asia and Pacific	7	9	10	38	35
South Asia	13	16	10	20	41
Europe	11	17	9	34	34
Middle East and North Africa	17	6	6	33	37
Latin America and Caribbean	12	13	7	31	35
Severely indebted	15	11	9	31	35
High-income economies	9	11	7	34	39
(OECD members)	(9)	(11)	(8)	(34)	(39)
World	9	11	8	34	39

Source: World Bank (1992) *World Development Report 1992*, New York: Oxford University Press, Table 15, pp. 246–7

Table 2.3 Structure of merchandise exports, 1990 (percentage share of merchandise exports – weighted averages)

	Fuels, minerals and metals	Other primary commodities	Machinery and transport equipment	Other manufactures	Textiles and clothing
Low-income economies	27	20	9	45	21
(China and India)	(10)	(17)	(15)	(58)	(26)
Middle-income economies	32	20	17	33	9
Lower-middle-income	32	30	11	27	9
Upper-middle-income	32	13	20	37	9
Sub-Saharan Africa	63	29	1	7	1
East Asia and Pacific	13	18	22	47	19
South Asia	6	24	5	65	33
Europe	9	16	27	47	16
Middle East and North Africa	75	12	1	15	4
Latin America and Caribbean	38	29	11	21	3
Severely indebted	42	22	14	22	5
High-income economies	8	11	42	40	5
(OECD members)	(7)	(12)	(42)	(39)	(4)
World	12	13	36	39	6

Source: World Bank (1992) *World Development Report 1992*, New York: Oxford University Press, Table 16, pp. 248–9

accounting for 36 and 39 per cent respectively. Figures for the group of countries representing East Asia and the Pacific and South Asia show that 47 and 65 per cent respectively of their export trade is in other manufactured goods, whereas in sub-Saharan Africa, 63 per cent is in fuel, mineral and metal exports and 29 per cent in primary commodities. These patterns together provide some explanation for the differences in income and development status of the various economies under discussion.

Direction of trade

To appreciate the direction of trade, it would be necessary to examine the destination of exports. However, as data are unavailable in a disaggregated form, and as it is impractical to analyse the direction of trade country by country, data would be required that show the origin and destination of exports by major groups of countries. Rather than continue the analysis according to income classes, it may be more useful if the data were aggregated according to broad trading blocs. Such an approach has two advantages. First, it would allow gross patterns to be detected so as to simplify description and explanation and, second, the analysis would be more appropriately discussed in the chapter on trade reform, regional trade groupings and the impact of GATT. The country composition of the regions are given as notes in Table 2.4. Table 2.4 and Figure 2.2 show the origin and destination of exports in 1979 and 1989. The data are obtained from the International Monetary Fund *Direction of Trade Statistics Yearbook, 1983–9*. An analysis of the data shows the high degree of intra-regional trade with the EU recording nearly 60 per cent, NAFTA about 40 per cent and ASEAN nearly 18 per cent.

A further observation is that the EU and NAFTA together account for over half of the origin of global trade, while Japan, the newly industrializing economies (NIEs) and ASEAN together account for about one-fifth of this trade. However, there is no symmetry in the reciprocal trade pattern in that while the EU and NAFTA are the destination of about 60 per cent of all exports, the latter group comprising Japan, NIEs and ASEAN are the destination of only 16 per cent of all exports. A detailed description of trade flows may be obtained by examining trade patterns on the basis of major regional groupings.

Table 2.4 Origin of exports classified by country groupings* and destination of exports, 1989 (US$ m.)

Origin of exports	Destination of exports										
	Australia	NAFTA	EU	EFTA	East 3	Japan	ASEAN	NICs	NZ	Other	Total
Australia	–	4,470	5,088	654	180	9,761	3,549	3,103	1,890	8,342	37,037
NAFTA	9,302	205,342	100,312	12,946	683	53,765	17,365	36,518	1,269	71,730	509,232
EU	8,795	100,753	677,824	118,185	10,220	23,216	14,807	18,725	1,419	159,708	1,133,652
EFTA	1,622	17,113	105,878	25,870	2,897	4,485	2,641	3,731	211	22,319	186,767
East 3	82	922	9,081	3,023	2,999	362	242	66	7	13,466	30,250
Japan	7,773	102,629	47,986	7,982	362	–	25,858	43,312	1,340	37,355	274,597
ASEAN	2,771	27,131	17,625	1,413	223	22,970	21,611	11,976	314	15,435	121,469
NICs	3,947	71,176	28,965	4,647	141	26,927	15,101	17,291	631	34,671	203,497
NZ	1,675	1,444	1,523	48	20	1,538	488	614	–	1,499	8,849
Other	2,856	74,946	129,959	14,952	11,044	54,911	13,716	32,847	2,387	–	337,618
Total	38,823	605,926	1,124,241	189,720	28,769	197,935	115,378	168,183	9,468	364,525	2,842,968

Source: IMF (1990) *Direction of Trade Statistics Yearbook, 1977–83, 1983–89*, New York: IMF
Note: *Country groupings: NAFTA: Canada, Mexico, US; EU: twelve current member countries; EFTA: Iceland, Finland, Norway, Sweden; East 3: Czech Republic, Hungary, Poland; ASEAN: Brunei, Indonesia, Malaysia, Philippines, Singapore, Thailand; NICs: Hong Kong, Republic of Korea, Taiwan.

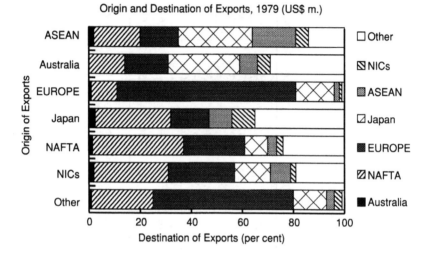

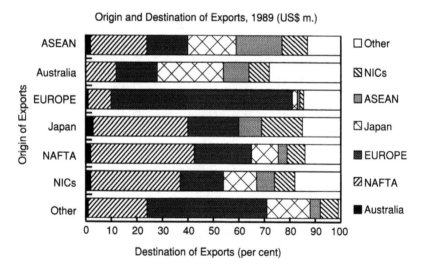

Figure 2.2 Origin and destination of exports, 1979 and 1989
Source: International Monetary Fund, *Direction of Trade Statistics Yearbook, 1977–83; 1983–89*, New York: IMF

Patterns of trade

Canada and the US

Canada and the US have the single largest trading relationship in the world and are each other's largest partner. In 1990 Canada's merchandise exports to the US were worth US$95 billion, 73 per cent of its total exports, while US merchandise exports to Canada totalled US$83 billion, 21 per cent of total US exports.

Mexico

Mexico is the US's third largest trading partner. In 1990, US merchandise exports to Mexico totalled US$28 billion, 7 per cent of US total exports. Mexico's merchandise exports to the US were worth US$22 billion representing 73 per cent of Mexico's exports. A high proportion of trade between the US and Canada and Mexico is in the form of intrafirm trade demonstrating the interdependence of investment and trade.

The asymmetrical pattern of trade dependence in the NAFTA bloc is repeated with imports. Mexico and Canada obtained 70 per cent and 63 per cent respectively of their total imports from the US in 1990. Meanwhile the US obtained 18 per cent of its imports from Canada and 6 per cent from Mexico. Clearly the US trade is very important for both Canada and Mexico, even though the US has other important global trading partners such as the EU, Japan and other Asian countries.

The Canada–Mexico relationship is symmetrically small; in 1990, less than 1 per cent of Canada's exports went to Mexico and about 2 per cent of Mexico's exports went to Canada. Imports from each country were similarly small. This is not surprising since the US is the dominant partner for almost every substantial trading country in the Americas; even distant Argentina does much more trade with the US than with its neighbour, Brazil.

Europe

Data on trade in goods suggest that the level of intra-European trade increased substantially during the 1980s. Exports from European countries to other European countries rose from 63 per cent of total exports in 1979 to 71 per cent in 1989. Imports similarly rose from 68 per cent of total imports in 1979 to 70 per cent in 1989. Within Europe, the EU is the most important trading partner, both for member states and for other western European countries. In 1979, the EU was the destination for 53 per cent of exports from EFTA countries and the source of 59 per cent of EFTA country imports. By 1989, the percentage of EFTA country exports going to the EU had risen to 57 per cent while the percentage of imports had declined marginally to 58 per cent. Trade between the EU and EFTA countries was 8 per cent of world trade – almost as large as the EU's trade with the NAFTA countries and Japan combined. By contrast, trade between EFTA countries is much less important, accounting for slightly more than 14 per cent of EFTA exports. Intra-community trade dominates the trade of EU member states, with nearly 60 per cent of member states' exports in 1989 going to other member states, an increase of nearly 9 per cent since 1979.

Asia

The East Asian economies comprising Japan, the NIEs, the ASEAN countries and China have shifted the balance of global economic power. Economic reforms based on export-oriented production and opening the economies to foreign trade and investment has resulted in growing shares of world trade. Between 1979 and 1989 the region's share of world merchandise exports grew from 12 per cent to 21 per cent and imports grew at a similar rate with East Asia's share of world imports about one-fifth (note that the data exclude China).

Any discussion of Asian trade cannot ignore the importance of the US market. However, with the appreciation of the yen, the Korean won and the Taiwan dollar against the US dollar and slower growth in US demand for imports, there has been a decline in the growth of Asian exports to the US. The share of Japanese exports to the US fell from 37 per cent in 1985 to 32 per cent in 1990. The same pattern applies to the NIEs, with Hong Kong and Taiwan experiencing the largest decrease in dependence on US markets from 44 per cent and 48 per cent in 1985 to 29 per cent and 32 per cent in 1990 respectively. China is the exception; here the US market is becoming increasingly important. The share of exports from ASEAN and NIEs going to Japan has decreased over the 1979–89 period as these economies have become increasingly internationalized. ASEAN merchandise exports to Japan were of similar magnitude to those going to the US (US$27 billion) in 1990.

Japan has increased the share of manufactures in its imports from 27 per cent in 1984 to 47 per cent in 1990. However, while the main sources of manufacturing imports are from the US and the EU (30 per cent share each in 1990), Korea, Taiwan, Thailand, Indonesia and Australia are all increasing their manufacturing exports to Japan by four- or five-fold in the 1980s. The NIEs are also increasing their imports of manufactured goods as their economies restructure.

How the above patterns of trade have evolved may be traced to various factors such as historical, factor endowments, political and economic structures and trade policies. It is to these variables that we turn next to observe how developing countries have attempted to enlarge trade in a global environment using various development strategies.

Development strategies

The trade patterns shown above prompt a re-evaluation of development strategies and trade policies for developing countries. The discussions and re-evaluation must consider both the efforts that produce the quantitative increase of growth such as export products and industrial goods and the variables that are responsible for the growth itself. In the present context, growth for any economy implies an expansion in the production of goods and services for home consumption and for export. R. Gilpin in his book, *The Political Economy of International Relations* (1987: 171), has noted that where growth results in an expansion of trade, a number of positive and related outcomes ensue. Apart from technological diffusion, there is the very important demand or so-called Keynesian effect on the economy that through the operation of the 'multiplier' stimulates further economic growth and the overall efficiency of the economy. Then there are other benefits for industrial firms because, as trade increases, the size of the market enlarges thereby promoting economies of scale, greater returns on investments while stimulating the overall level of economic activity as a whole. More goods in the marketplace means a larger range of consumer choice and a reduction in the costs of inputs such as raw materials as well as manufactured components which combine to lower the cost of production. This export-led growth strategy may also be used to acquire the necessary imports to stimulate further growth. This viewpoint is a useful one since it suggests the close relationship between development strategy and trade policy. However, what is an appropriate strategy or policy is a contentious issue since it will be difficult to deny that whatever commercial policies that a developing country pursues may have a significant impact on the pattern and pace of economic growth and development.

There are a number of production and trade policies that may be adopted from autarchic self-sufficient regimes to import substituting industrialization (ISI) and export-oriented policies. An inward-oriented strategy such as ISI is one in which trade and industrial incentives are biased in favour of production for the domestic rather than the export market. In the 1950s and 1960s import substitution was the dominant strategy and used a wide range of protectionist policy instruments to encourage new (and expand existing) industries. In the process, both tariff and non-tariff barriers were used extensively, with many policy-makers preferring the latter, in the form of import licensing, quantitative restrictions and local-content requirements. The Philippines, for instance, reduced its imports of consumer goods from 30.9 per cent to 4.7 per cent of total supplies between 1948 and 1965. In these strategies specific trade policies provided the preferred route to development. At the same time the trade policy itself constituted part of the overall development planning process, tailored to fit conditions and potenti-alities of particular countries. Trade policies were also geared to meet-ing objectives and priorities set in development plans such as income distribution, unemployment and poverty.

Import substituting industrialization (ISI)

To counteract the imbalance of their overseas trade and to overcome the adverse terms of trade when trading primary commodities for industrial goods, many developing countries have adopted ISI. A contraction in exports brought about by the development of substitute goods overseas together with greater competition in those markets and the need to retain demand in home markets encouraged many developing countries to consider ISI policies and strategies. Home-produced goods, moreover, could prevent a rise in domestic demand from leaking too greatly abroad, and at the same time conserve valuable foreign exchange. These efforts would counteract the trend towards unfavourable balance of payments while providing fuller industrial employment. However, ISI demanded an abandonment of free trade policies and a free floating monetary system, and thus has been labelled inward-looking industrial-ization by some observers. Economic rationalists have also argued that a reduction in imports would increase employment and raise incomes and stimulate further industrialization. Here the notion of comparative advantage becomes irrelevant because the general protection accorded to industries will stimulate general industrialization regardless of the costs of local production.

The record for ISI around the developing world has been unspectacular. It seems that there is an inverse correlation between the degree of 'success' in ISI and the overall growth rate of the economy. Forced import substitution, as induced by high tariff barriers, may result in the growth of inefficient industries and a misallocation of scarce resources. This means that high-cost industries supply the domestic market and that these same industries will be unable to compete in the world market. These 'protected' industries demand ever-increasing amounts of government concessions and thereby resist having to test themselves in the rigours of a free and competitive market.

Bela Balassa, writing in the *American Economic Review* (1971), has observed that while the protection of the manufacturing sector may permit growth at the early stages of ISI, it will eventually have adverse consequences for economic growth for a number of reasons:

1 The discrimination among industries will bar specialization according to comparative advantage.
2 The high protection of domestic industries will induce the establishment of high-cost import substituting activities.
3 There will be the creation of a bias against exports.
4 The absence of foreign competition will give little incentive for technical progress in small protected markets.

The solution seems to be that trade policies should aim at encouraging exports and promoting greater efficiency in the use of resources. The distribution of income should be more equitable in combination with creating more employment in industries and primary production. The suggestion therefore is that an export-oriented strategy may be more appropriate for most developing countries, but this should be supported by economic policy that helps buttress those industries for which unexploited import substitution possibilities exist.

Export-oriented strategies

Jagdish Bhagwati and Anne Krueger, writing in the *American Economic Record* (1973), have argued that export promotion may indeed be a superior strategy to import substitution. First, the costs of export promotion are more visible to policy-makers than those of import substitution.

Second, there is consíderable evidence in individual country studies to show that direct intervention may be more costly than is generally recognized. Third, the monopoly positions assumed by ISI have contributed to low productivity growth in newly established manufacturing industries in developing countries. Finally, an export-oriented growth strategy is better suited to achieving whatever economies of scale are present than is an import substituting strategy in which firms are generally limited in their scope by the size of their domestic markets.

Most developing countries are now seeking to adopt an export-oriented strategy either by increasing local processing of indigenous raw materials or by export-oriented industrialization. Such a strategy also requires a parallel restructuring of imports away from manufactured and consumer goods from developed countries towards the import of capital and intermediate goods. Although this restructuring may have an adverse effect on the balance of payments in the short term, the argument is that these structural adjustments may be beneficial in the long term.

Outward-oriented export promotion strategies in theory do not discriminate between production for the domestic market and for exports, or between purchase of domestic goods and foreign goods. Some of the policy instruments which have been employed include financial and fiscal incentives for export producers, export guarantees and letters of credit, the setting of export targets and the establishment of export processing zones and/or free trade zones. The first group of countries to pursue this path include the Four Dragons – Singapore, Hong Kong, Taiwan and South Korea – later followed by other semi-industrialized and newly industrialized countries (NICs) of ASEAN. However, the precise form of export orientation has varied according to specific factor endowments in each country. Countries with favourable primary resource endowments have attempted to change from mainly primary exporters to manufacturing exporters through local processing. In other countries, manufactured export capacity has been based on abundant cheap labour supply combined with investments by MNCs.

The success of the Asian Four Dragons and other NICs in generating exceptional growth rates from manufactured exports and output has become a model for most developing countries to emulate. International development agencies like the World Bank are strongly committed to encouraging developing countries to adopt outward- rather than inward-oriented strategies. A World Bank study in 1987 of 41 developing countries covering the period 1963–85 has shown that if countries are placed along a continuum of strongly outwardly-oriented at one end and strongly inwardly-oriented at the other, economic performance including GNP growth tends to decline from outward- to inward-orientation. The study suggests that strongly inwardly-oriented economies fared badly. While the results of the study are highly persuasive, they fail to take into account the relative poverty of the inwardly-oriented group of countries when compared to the other group of outwardly-oriented countries, a characteristic which is even more important than the lack of economic performance. It may be that poorer countries have greater difficulty than relatively richer countries to progress up the ladder of development.

On another analysis, trade orientation is positively related to levels of per capita income and outward orientation is more closely associated with a balanced export structure. Where the export structure is dependent on a narrow range of primary products, such economies urgently need to diversify their external trade to remove the threats emanating from the vagaries of international trade.

Although the potential benefits of an outward-oriented strategy are immense, it is appropriate to conclude this section with a cautionary note expressed elegantly by Michael Todaro:

> An export oriented strategy of growth, particularly when a large proportion of export earnings accrue to foreigners may not only bias the structure of the economy in the wrong directions (by not catering to the real needs of local people) but may also reinforce the internal and external dualistic and inegalitarian character of that growth. Therefore, the fact that trade may promote expanded export earnings, even increased output levels, does not mean that it is a desirable strategy for economic and social development. It all depends on the nature of the export sector, the distribution of its benefits, and its linkages with the rest of the economy.
>
> (1981: 386)

The record of import substitution and export-oriented strategies has been unreliable in some cases. Internally, countries would need to have in place basic infrastructural facilities in order to compete in world markets. However, countries have little or no control over external influences, no matter how prepared. For this some degree of serendipity and luck would be required. Case study B demonstrates the plight of three South American countries.

Case study B

South America's free market approach

During the 1950s and 1960s Argentina, Chile and Uruguay favoured introspective import substitution industrialization, reliant on extensive government intervention, anti-export bias and protection. In the 1970s, severe inflation and balance of payments crisis forced a switch towards a more open economy, free currency movements and *laissez-faire* policies. Controls on prices, trade, interest rates and capital flows were reduced. Government expenditure was cut and an active programme of privatization put in place. All three economies experienced initial success with manufacturing expanding, inflation brought under control and capital inflows combining with exports to redress the balance of payments.

Chile went furthest in its attempts at liberalization, deregulating domestic prices, reducing tariffs and non-tariff barriers and practically eliminating subsidies. During 1975–9 investments flowed to export manufactures and non-traditional exports grew by 32 per cent per year.

In Argentina import-competing manufacturers were still protected while in Uruguay exports were positively encouraged by new financial incentives. Manufacturers' share of exports in Argentina rose from 10 per cent through 1965–7 to 16 per cent in 1968–70 and peaked at 29 per cent in 1980–2. In Uruguay, manufactures grew dramatically from 26 per cent of exports in 1965–7 to a peak of 45 per cent in 1977–9 before falling back. (But these are small compared to South Korea's manufacturing record which rose from 64 per cent of exports in 1965–7 to 92 per cent in 1983–5.) However, these improvements collapsed in the 1980s in all three countries. Exchange rates unintentionally discriminated against exporters, tariff protection and incentives were removed, corporations used overseas debt for risky but high-profit arbitrage rather than productive investment, and foreign investment flooded into the non-traded service sector and property and away from manufacturing. Arbitrage is a business operation involving the purchase of foreign exchange, gold, financial securities or commodities in one market and their almost simultaneous sale in another market in order to profit from price differentials existing between the

Case study B (*continued*)

markets. Import substitution firms even became retailers of cheaper imported goods. By 1983–5 the share of manufacturing in exports had fallen back by a third in Chile and Argentina and by a quarter in Uruguay. In contrast, manufacturers' share of exports climbed in Brazil, Hong Kong, Taiwan and South Korea. Despite differences in policies and in the degree and nature of deregulation in Argentina, Chile and Uruguay the familiar boom–bust cycle is evident. These suggest difficulties:

- of sustaining growth in a climate of partial deregulation;
- of fiscal policies aimed at stabilizing the economy but which run counter to export-led growth;
- of the destabilizing effects of fickle foreign capital inflow; and
- of coordinating government policies to stimulate growth.

Source: D. C. Kemp (1993) 'Can Australia and other Third World countries emulate Korea's experience with economic development?', in Robert Leach (ed.) *National Strategies for Australasian Countries: the Impact of the Asian/Pacific Economy*, Brisbane: Queensland University of Technology, pp. 96–112.

Direct foreign investments (DFI)

For NICs direct foreign investment (DFI) may yet be another strategy assisting overall development. This is because an export-oriented strategy carries with it an implicit assumption of 'openness', one in which a country's goods may find a market in other countries in the same way that foreign goods may find a market within it. Openness also means the free flow of capital, of people and of knowledge – transmitting technological change and generating economic growth across nations. With technological change in the form of technology transfer, there will be increasing global competition thereby raising the demand for more and newer technology. This supply of new technology will depend to a large extent on the degree to which industrial countries are integrated in the global economy and the place or position of the developing country within the scheme of things. In Hong Kong and Mexico, for instance, the presence of foreign firms has increased the diffusion of technology. In

Brazil, too, a large share of manufactured exports originates from firms with foreign investment. It is here that the role of the multinational corporation (MNC) comes to the fore since the more established the MNC is in a developing country, the more likely will there be a fuller integration with the local economy. This flow of capital in the form of DFI and of skilled workers provides an important source of technology transfer and diffusion of innovation as well as fostering export growth.

Inflows of foreign capital can finance domestic investment and help economies begin the path to development. However, gains from DFI depend on government policies and political attitudes since foreign investment in a protected market may hinder development rather than promote it. For example, in the Côte d'Ivoire (Ivory Coast) the selective protection and subsidies given to a multinational textile firm led to inefficient production. On the other hand, joint ventures in Morocco (phosphates) and Venezuela (petroleum and aluminium) have proved more successful than merely restricting foreign ownership. Joint ventures in Venezuela export a higher share of the total output than domestically owned firms.

Labour mobility may be another avenue for reducing the disparity of incomes world-wide through net remittances to their home countries from labour migrants. Net remittances from migrants in France, Germany, Kuwait and Saudi Arabia are often quite high. About 10 to 50 per cent of every dollar earned is sent to the migrant worker's home country. On the negative side, such movements may be viewed as a brain drain of skilled personnel to the detriment of the country of origin. Bangladesh's shortage of professional workers is attributed to emigration of these workers. The high proportion of students from South Korea (63 per cent), Jordan (49 per cent) and Greece (33 per cent) remaining in the US after completing their studies is a further example.

The mere presence of foreign firms may increase technological diffusion as well as improve the efficiency of local firms. Different forms of management and marketing skills may be as important as the transfers of the product or the processes and technologies associated with manufacturing the product. Industrialists in Bali, Indonesia and Taiwan have benefited from their association with foreign firms and have entered world markets without needing to invest in market research and without the usual entry costs. Other countries such as Malta, Mauritius and Singapore have exploited their linkages with foreign firms that have global markets.

The flow of DFI is likely to grow and become more widespread as a response to policy changes. However, DFI has been and will continue to be concentrated in middle-income countries that have well-developed infrastructures. According to the World Bank (1991: 24) in 1989, about 70 per cent of DFI flows to developing countries came from Japan (18 per cent), UK (20 per cent) and US (32 per cent). Just 20 developing economies in Asia and Latin America were beneficiaries of such investments and accounted for 90 per cent of the net flows between 1981 and 1990. The economic reconstruction of eastern Europe and the Commonwealth of Independent States will undoubtedly increase competition for DFI in the future.

In 1980 DFI was estimated at US$9.1 billion and grew to US$22 billion in 1989. The projection is for DFI to grow to US$35 billion by 1995. Between 1970 and 1989 DFI grew at an average rate of 6 per cent annually with a steady upward trend in the 1970s dropping slightly between 1981 and 1986. DFI increased about 12 per cent a year between 1970 and 1989 for Asia compared with 3 per cent in Latin America and a decrease in Africa. The percentage shares of DFI as net flows to developing countries was only 11 per cent in 1980 and grew to 35.3 per cent in 1989. It is projected to remain at nearly the same share in 1995.

There is uncertainty, though, for enhanced flows of DFI to developing countries in the 1990s and beyond. One study estimates that the share of developing countries in global foreign investment flows declined from 26 to 21 per cent in the 1980s. Moreover, DFI is highly concentrated, with 15 countries attracting 75 per cent of all investment in the 1980s.

In summary, DFI is a potentially important source of capital to supplement domestic investment, technology transfer and employment generation. However, the regional integration of Europe may yet prove more attractive to foreign investment, thereby discouraging flows to developing countries. Increasing protectionism diverts DFI from other destinations and makes developing countries less attractive locations for export-oriented foreign investment.

Conclusion

The patterns of trade and the strategies and policies discussed above suggest that government intervention will produce higher productivity growth where there are no undue price distortions. While there are considerable variations in countries' experiences, successful market intervention will only result if such interventions are 'market friendly'. This means that governments should intervene reluctantly, apply checks and balances in all policies pursued and to intervene openly. The record shows that successful intervention preserves incentives for technical change by maintaining international and domestic competition and by imposing preferential requirements in return for credit subsidies, import protection and the restriction on domestic entry.

Whether this prescription for government intervention will work for all countries is a moot point. However, the ingredients for sustained growth and ultimately economic development have been identified. Economies are increasingly seeking to integrate either geographically with economies which are near neighbours or with MNCs to bring about economic development. The aim is to 'widen' the development process in a spatial sense, and also to 'deepen' the economic structure. The ultimate aim is for an economy whose trade structure, trade policies and strategies are sustainable in a global economic climate which may oscillate from prosperity to recession and back to prosperity again.

The next chapter examines one particular aspect of the growth and development dilemma – that of aid. As a result of adverse terms of trade, the worsening balance of payments of many low-income economies and the growing indebtedness among these economies, there is a need to analyse financial flows in the form of aid and investments. Also the relationship between trade and aid will be explored in the context of the politics of aid.

Key ideas

1 International trade and economic development are closely linked. Such a relationship may be explained by a spatial model where economic production, interactions and physical links manifest themselves in space.

2 Growing productivity brought about by technological progress has been a significant contributor to the growth in world trade.

3 Developed countries with only one-fifth of the world population produced four-fifths of world output and four-fifths of world trade and almost all exports of capital and technology.

4 Development strategies and trade policies must be tailored to suit particular countries.

5 Import substituting industrialization (ISI) is a means of retaining demand in home markets and of expanding exports. But such policies often lead to protectionist policies which may be counterproductive to a dynamic economy.

6 Export promotion strategies encourage efficient production where

there is no discrimination between the domestic or the export market. The Four Dragons have successfully generated exceptional growth rates in their economies by using export-oriented strategies.

7 Direct foreign investment (DFI) is an important source of capital to supplement domestic investment, technology transfer and employment generation.

3
Aid and development

Introduction

This chapter is about the role of aid, financial flows and the debt crisis confronting Third World countries in their path towards development. The advent of international aid divided the globe into various 'worlds'. Because of the need to label the group of countries that either did not belong in the developed industrialized First World or to the socialist Second World, the 'Third World' was coined as a generic term to encompass all developing countries. Regrettably, we also have a 'Fourth World' where some countries appear to have little or no hope of development in the near future. Apart from development status, some countries in the Third World are classified as being severely in debt. For example, in terms of the World Development Indicators of the World Bank 15 countries were deemed to have encountered severe debt-servicing difficulties. These countries are said to be experiencing critical problems in three of the four key ratios. These include a debt to gross national product (GNP) ratio of over 50 per cent, a debt to exports of goods and all services ratio above 275 per cent, accrued debt to exports ratio of more than 30 per cent, and accrued interest to exports ratio greater than 20 per cent. The 15 severely indebted countries are Algeria, Argentina, Bolivia, Brazil, Bulgaria, Congo, Côte d'Ivoire (formerly Ivory Coast), Ecuador, Mexico, Morocco, Nicaragua, Peru, Poland, Syrian Arab Republic and Venezuela. Eight of these countries

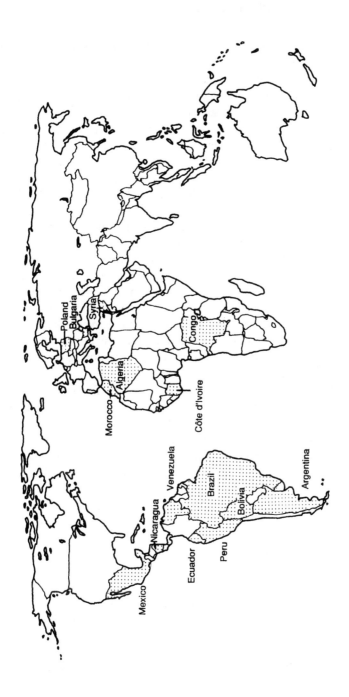

Figure 3.1 Fifteen severely indebted nations

are in Latin and South America, four in Africa, two in Europe and one in the Middle East (see Figure 3.1).

The suggestion here is that a debt crisis is prevalent and touches all kinds of economic structures. As a result of an interplay of complex events, the debt problem unfolded in the 1970s when many developing countries borrowed money to increase consumption, to invest in doubtful projects and to finance imported oil. The World Bank (1991) estimated that during the decade (1970–80) the volume of international bank lending increased by nearly 800 per cent to about US$800 billion. Many lending agencies, both institutional and commercial, did little detailed investigation on how the loans were being used but relied on sovereign assurances. The low productivity of the investments culminated in a debt crisis that coincided with the world recession of the mid-1970s. The resulting high interest rates, and deteriorating terms of trade caused severe debt-servicing problems forcing many countries to re-negotiate and restructure loan agreements. Some countries, like Mexico in 1982, were forced to declare a debt moratorium, while others experienced falling investments and found it difficult to recover in time to take advantage of a more favourable global environment in the 1980s. The cumulative result was negative, followed by very slow growth during the decade of the 1980s.

This chapter examines the role of aid and financial flows in alleviating the debt crisis and its relationship to economic growth. In the first section we examine the term aid in all its forms and how this bears upon an understanding and analysis of the relationships between aid and development, and between aid and trade. The second section provides a statistical summary of aid flows by donor and recipient. The third section identifies the roles of the major aid agencies, such as the World Bank and the activities of non-government organizations (NGOs) in delivering the aid programme. The fourth section examines the effects of aid on development together with the growing debt crisis in developing countries since it seems that either the aid programme is failing to produce the desired results of development or that the aid has been channelled to sectors of an economy with low productivity thereby undermining debt-servicing. The final part of this chapter addresses the vexed questions of the politics of aid in terms of why donors give and why recipients accept aid, and the view that aid is imperialism of a different form especially when aid is given in return for trade rather than generating the ability to trade through aid. This final issue is also taken up in the next chapter when discussing the transfer of technology to developing countries.

Aid: definitions and forms of . . .

In simple terms, aid refers to the flow of resources from developed countries to developing countries. Such resources may be in the form of finance or in the form of goods and/or technical services, the cost of providing these being usually covered by donor countries. Aid is bilateral when it involves government to government transfers, whereas it is multilateral when international organizations provide such assistance. Bilateral aid, which accounted for three-quarters of total aid between 1980 and 1988, includes aid from member countries of the Development Assistance Committee (DAC) of the Organization for Economic Cooperation and Development (OECD). Aid from international organizations such as the various Development Banks, the European Community, the Organization of Petroleum Exporting Countries (OPEC) and other United Nations agencies are examples of multilateral assistance. World-wide, private voluntary agencies may also provide aid and these bodies have generally been labelled non-government organizations (NGOs). Private investments and loans by commercial banks to developing countries are included as part of international trade and are rarely included as part of aid, although there is a strong connection between the two.

A fundamental rationale for economic aid is that a country needs help to reach a stage of sustained economic growth. When that stage has been reached economic aid can be reduced and eventually cut off. The aid may be given in a number of ways. Capital transfers take the form of cash or kind and can be given as either grants or loans. While grants may be given outright, recipient countries have to repay loans, albeit on concessional terms. Such concessional terms as a grace period of five years where no repayment will be required and a repayment period over 20 years or less at low interest rates are quite usual for these so-called 'soft' loans. Technical assistance and training represent a transfer of technical equipment and/or personnel. Such transfers are most important at the initial stages of economic development because, as technical know-how in the use of modern equipment increases, the country will be more able to absorb further inputs of capital. While the distinctions made above are fairly straightforward, whether or not military assistance in the form of either material or training has an economic component is

Plate 3.1 Food aid: a shipment of rice as part of the World Food for Africa Progamme from Australia. Note the use of conventional loading methods because of the lack of modern handling facilities at the destination ports
Photo: Australian Department of Foreign Affairs and Trade

not so clear-cut. As witnessed in the American involvement in South Korea, Taiwan and South Vietnam, technical advisers together with budgetary support were a feature of the military assistance.

Finally, one form of aid which is distinctive in its own right is disaster and emergency relief for natural and human-induced phenomena. Refugees from natural disasters and from civil conflict have been assisted by relief organizations, governments, the UN and others, for example, the International Committee of the Red Cross (Red Crescent in Islamic countries), *Médecine sans Frontières* and the UN High Commissioner for Refugees (UNHCR). Plates 3.1 to 3.4 illustrate graphically food aid as a result of crop failures and natural disasters, appropriate aid in the form of village water supply and even as technical assistance and facilities. Here the aid is in the form of medical supplies and the establishment of a basic health infrastructure to help ensure that the facility continues to provide services over a long period. Finally, human misery brought about by natural disasters calls for assistance of a special nature, for instance, disaster relief to help refugees. Here the role

Plate 3.2 Appropriate aid: village water supply in the Philippines. Development projects that supply clean water for drinking and washing are essential if primary health care and sanitation are to be maintained and diseases avoided
Photo: Australian Department of Foreign Affairs and Trade

Plate 3.3 Aid in the form of technical assistance and facilities. Picture shows a UNICEF Health Station in Babon, North Samar, the Philippines
Photo: Australian Department of Foreign Affairs and Trade

Plate 3.4 Disaster relief: assistance to refugees from natural disasters. Photo shows the interior of a Rohingya hut in Bangladesh. The minimal needs of food, clean water and shelter provide some hope of relief
Photo: Australian Department of Foreign Affairs and Trade

of pure altruism is difficult to measure – being perhaps the main factor behind the work of voluntary agencies and many donor countries.

The transfer of resources can also take other forms sometimes merging with other activities. For example, economic aid may take the form of preferential tariffs, that is, developed countries offer Third World countries a market for manufactured goods. This permits the sale of developing country goods at higher prices due to tariff reductions. There is a gain for the developing country and a 'loss' for the developed country, in effect a real resource transfer to the former. On the other hand, some transfers of capital to developing countries arise from private investors. These represent normal commercial transactions and therefore are not a form of foreign assistance even though developing countries may benefit from them.

A strict definition of economic aid thus is any flow of capital to developing countries where its objective is non-commercial from the point of view of the donor country, and is characterized by concessional terms in which the interest rate and repayment period for the borrowed capital is less onerous than commercial terms. Such a definition will also include military aid since it fulfils both criteria. However, in practice such aid is recorded separately for conceptual and reporting purposes. An underlying motive for such economic aid is that the transfer of capital is on developmental and/or income distribution grounds.

Foreign aid can be 'tied' either by source or by project. Aid tied at source will require the loans and/or grants to be spent on the purchase of donor country goods or services. Tied aid project funds cannot be spent on raw materials and intermediate products. The result is that the value of tied aid to the recipient country is reduced since it will have to be spent in a specific way which is likely to be more expensive. Moreover, having to import capital-intensive equipment may promote higher unemployment in the recipient country as such equipment replaces labour. A further interpretation of 'tied' aid is that because recipient countries have to purchase goods from donor countries, the higher prices aid recipients have to pay when compared to buying in other markets represent the payment of a hidden interest charge – ironically an element of finance which developing countries had hoped to avoid in the first place when seeking foreign aid!

This theme of tied aid may be illustrated by recent experiences in Malaysia. Britain promised project aid to the value of US$627 million to help finance the building of the Pergau 600 mW hydroelectric dam in the northern state of Kelantan. This dam involves the largest cash sum ever provided for a single scheme under Britain's Overseas Development Administration's (ODA) Aid and Trade Provision programme. The aid represents more than half the costs of the project which is scheduled for completion in the middle of 1995. The aid money is in the form of 'soft loans', that is, it involves interest subsidies paid by the British government that help participating banks make up the difference between market rates and concessional rates granted to foreign governments. As a result Malaysia was only required to pay interest at 0.809 per cent over a repayment period of 14 years. Such generosity, however, sparked an audit by the British National Audit Office which released a report in October 1993. It is from this report that we learn about the nature of the 'tied aid'. The report showed that the Conservative government proceeded with the provision of US$350 million in 1991 to finance the Pergau project even though aid officials opposed the project as 'uneconomical' and a waste of money. The report also gave evidence of efforts to use foreign aid to win contracts for British companies in Malaysia. It helps explain how the British have become one of the chief recipients of Malaysian contracts since 1988.

In fact, two leading British firms, Balfour Beatty and Cementation International, are building the Pergau dam in partnership with a local company, Kerjaya Binaan. The report also shows that fears of strained Anglo-Malaysian relations and the need to secure contracts for British companies provided the impetus for the British government decision to support the Pergau project. The aid question therefore needs to be approached with caution and there is more to it than meets the eye.

Food grants are a form of non-financial aid. Together with technical assistance such grants either release resources for other purposes or provide resources that are otherwise unavailable. For example, foreign exchange which formerly was used to pay interest on loans, or to import food, may now be reallocated to purchase machinery. The presence of foreign engineers, economists, teachers and other personnel as part of technical assistance can do more than mere monetary help. Food grants and disaster relief in the form of food to recipient countries has also been used as a method of reducing agricultural surpluses in donor countries – setting up a win–win situation.

The World Bank generally provides loans for development projects that aim to raise living standards in developing countries. One arm of the World Bank is the *Association Internationale de Développement* (AID) or the International Development Association (IDA) which provides credits to distinguish them from World Bank loans. Such credits are made to governments only, have a 10-year grace period and 50-year maturity with no interest. Thus, when the term official development assistance (ODA) is used it means that it consists of net disbursements of loans and grants made on concessional terms by official UN and other agencies.

Aid: donors and recipients

The World Bank aims to finance productive projects that promote economic development. Resulting from the Bretton Woods negotiations of July 1944, the Bank obtains funds from paid-in capital subscriptions of member countries, from bond flotations and from net earnings. Loans are made directly to governments or private enterprises with government guarantees for specific projects where private capital is not available on reasonable terms. The general policy is to lend only for the cost of imported material, equipment and services obtained from abroad and disbursements made directly to the buyer. Interest rates

Table 3.1 Official development assistance from OECD and OPEC members, 1980–90

	Total net flows (US$ m.)		As a percentage of donor GNP	
	1980	1990	1980	1990
OECD				
Ireland	30	57	0.16	0.16
New Zealand	72	95	0.33	0.23
Australia	667	955	0.48	0.34
United Kingdom	1,854	2,638	0.35	0.27
Italy	683	3,395	0.15	0.32
Netherlands	1,630	2,592	0.97	0.94
Belgium	595	889	0.50	0.45
Austria	178	394	0.23	0.25
France	4,162	9,380	0.63	0.79
Canada	1,075	2,470	0.43	0.44
Germany	3,567	6,320	0.44	0.42
Denmark	481	1,171	0.74	0.93
United States	7,138	11,394	0.27	0.21
Sweden	962	2,012	0.78	0.90
Finland	110	846	0.22	0.64
Norway	486	1,205	0.87	1.17
Japan	3,353	9,069	0.32	0.31
Switzerland	253	750	0.24	0.31
OECD total net flows	27,296	55,632	0.36	0.36
OPEC				
Nigeria	35	13	0.04	0.06
Qatar	277	1	4.16	0.02
Algeria	81	7	0.20	0.03
Venezuela	135	15	0.23	0.03
Iran, Islamic Republic	−72	2	−0.08	*
Libya	376	4	1.16	0.01
Iraq	864	55	2.36	*
Saudi Arabia	5,682	3,692	4.87	3.90
Kuwait	1,140	1,666	3.52	*
United Arab Emirates	1,118	888	4.06	2.65
Total OPEC[1]	9,636	6,343	1.85	*
Total OAPEC[2]	9,538	6,313	3.22	*

Source: World Bank (1992) *World Development Report 1992*. New York: Oxford University Press, Table 19, pp. 254–5
Notes: [1] OPEC – Organization of Petroleum Exporting Countries.
     [2] OAPEC – Organization of Arab Petroleum Exporting Countries (comprising Algeria, Iraq, Kuwait, Libya, Qatar, Saudi Arabia and United Arab Emirates).
   * Not available.

charged depend on the cost of borrowing to the Bank. As well as the IDA, the International Finance Corporation (IFC), an organization affiliated with but legally separate from the World Bank, aims to stimulate the economic development of member countries by providing capital for private enterprises when sufficient capital is unavailable. The IFC makes loans without government guarantees of repayment and is prevented from stipulating where its loans will be spent. Although the terms of the loans are negotiated in each case, maturity dates are usually between five and 15 years.

ODA is the term used to describe the disbursement of loans and grants. UN and other international loan agencies are usually members of the DAC of the OECD or members of OPEC who wish to promote economic development and welfare. The definition strictly excludes military assistance, but this is often blurred since some donor countries make no distinction in reporting the type of aid that has been provided. Table 3.1 shows ODA from OECD and OPEC members for 1980 and 1990 in millions of US dollars and as a percentage of donor GNP. To compare the percentage of donor GNP, a separate column is also shown for OECD and OPEC total bilateral flows.

Japan, US, France and the former West Germany together contributed US$36,163 million out of a total net flow of US$55,632 million from OECD countries in 1990. This amount compared to the US$18,220 million given in 1980 by the same four countries out of a total of US$27,296 million. While these four countries are the largest contributors of ODA, it is significant that their relative proportions have fallen from 66 per cent to 64 per cent over the same period. ODA as a share of the GNP of OECD countries remained stable at 0.36 per cent. However, the significance of this has been overshadowed by the increased contributions of all members of OECD on the one hand, and the decline in relative terms as a percentage of donor GNP on the other hand. Large increases over the 1980–90 period were recorded by Finland (+669 per cent), Italy (+397 per cent), Switzerland (+196 per cent) and Japan (+170 per cent), while the smallest increase was that of New Zealand with 32 per cent. Nine countries have shown an increase in the amount given as a percentage of donor GNP, while eight others have decreased in relative terms with one country unchanged (Ireland). Only Norway (1.2) gave slightly more than 1 per cent of its GNP as ODA. When ODA is shown as bilateral flows to low-income economies in 1990, the contributions of OECD countries were on average about one-third that of their total contributions.

The table also shows the contributions of members of OPEC. Here the pattern for 1980 and 1990 shows, with the exception of Nigeria, a relative decline for all countries listed. While Iran increased its contributions by 103 per cent and Kuwait by 46 per cent, the decline in contributions ranged from 99 per cent for Qatar to −21 per cent for the United Arab Emirates. The total value of OPEC contributions to ODA amounted to US$6,343 million in 1990, a decrease of 34 per cent from the 1980 figure of US$9,636 million. The data for 1990 also show that as a percentage of donor GNP, Saudi Arabia contributed 3.9 per cent of their GNP in development assistance and the United Arab Emirates 2.7 per cent; other OPEC member contributions ranged between 0.01 per cent (Libya) to 0.06 per cent (Nigeria). These fluctuations in the data demonstrate the unrest in the region surrounding OPEC countries both in terms of domestic and external political security and in terms of the stability of oil prices during the 1980s. The civil unrest in Nigeria, the rise of Islamic fundamentalism in Iran, the continuing Iran–Iraq conflict and the political problems in Libya all contributed to the fluctuations in aid contributions.

The data are perhaps symptomatic of the times. Recent trends in global development assistance and other net resource flows from developed countries have all shown negative growth. While ODA in the 1960s to the less developed countries increased by about 5 per cent per annum, in the 1970s the growth was about 30 per cent per annum. This large growth rate was shown to be unsustainable in the 1980s because of the onset of 'aid fatigue' and 'aid weariness' for both donors and recipients alike and the global economic recession. The trend for ODA flows shows some signs of stability given that the ODA as a share of GNP has remained at 0.36 per cent in the decade between 1980 and 1990. However, the UN has set a target of 0.7 per cent of GNP as the level of donations in development assistance for all countries. Nearly all countries have committed themselves to this target but not all have reached it, and although some have exceeded it, others are nowhere near the target.

Table 3.2 gives data on receipts of ODA for 1984 and 1990. Total disbursements of ODA from all sources increased from US$25,570 million in 1984 to US$49,070 million in 1990 representing a 92 per cent increase. The distribution of ODA showed that low-income economies were given about two-thirds of the total available, middle-income economies about a third and severely indebted countries a tenth. The breakdown of the data by major country groupings show that sub-

Saharan Africa received about US$16,810 million in 1990 (34 per cent), while East Asia and South Asia received about US$7,771 million and US$6,174 million respectively, slightly more than the assistance severely indebted countries received – US$4,660 million. Countries in the Middle East and North Africa, Latin America and the Caribbean received about US$9,680 million and US$5,380 million respectively. When computed as a percentage of the GNP in 1990 these receipts show that ODA made up 2.8 per cent in low-income countries, 1.6 per cent in lower-middle-income economies and only 0.4 per cent for the severely indebted countries. By far the largest ODA contributions as a percentage of GNP were to sub-Saharan Africa, 9.6 per cent, and 3.4 per cent to the Middle East and North Africa. This analysis reveals that ODA has been directed to those economies most in need. But the analysis also shows that such a statistical summary of the donors and recipients would be incomplete without a further discussion of the role of aid agencies and NGOs in the delivery of assistance.

Table 3.2 Official development assistance: receipts, 1984 and 1990

	Net disbursement of ODA from all sources (US$ m.)		*ODA per capita (US$)*	*ODA as % of GNP*
	1984	*1990*	*1990*	*1990*
Low-income economies	14,476	29,353	9.6	2.8
(China and India)	(2,471)	(3,662)	(1.8)	(0.6)
Middle-income economies	9,557	17,882	18.7	0.7
Lower-middle	7,730	14,365	26.0	1.6
Upper-middle	1,827	3,517	8.5	0.1
Sub-Saharan Africa	7,941	16,810	33.9	9.6
East Asia and Pacific	3,553	7,771	4.9	0.8
South Asia	4,585	6,174	5.4	1.6
Europe	376	1,420	14.1	0.4
Middle East and North Africa	4,506	9,680	37.8	3.4
Latin America and Caribbean	3,072	5,380	12.3	0.4
Severely indebted	2,379	4,660	11.4	0.4
High-income economies	*	*	*	*
Other	1,525	1,802	44.7	0.8
World	25,570	49,070	12.0	1.4

Source: World Bank (1992) *World Development Report 1992*, New York: Oxford University Press, Table 20, pp. 256–7
Note: * Not available.

Aid agencies and non-governmental organizations (NGOs)

Traditionally, development assistance has been organized and delivered by developed countries. More recently, however, there has been a steady increase in intra-Third World development assistance. The impetus for this has been provided by the enormous surplus of oil revenues accumulated during the mid-1970s by oil-producing Middle Eastern nations. While about US$54 billion was accumulated in 1974 which was available for 'recycling' to non-oil exporting countries, there was about US$100 billion available in the 1980s. For example, Saudi Arabia extended grants and credits to neighbouring Arab states like Egypt and Syria. While some of the assistance may be described as military and political support, much of it is for genuine economic development, for example, the creation of and contributions to the Arab Fund for Economic and Social Development and the Fund for African Development. Likewise, Venezuela has assisted Latin American countries in need, while Nigeria, despite domestic turmoil, initiated the Special African Aid Fund administered by the African Development Bank (AfDB). This special fund provides long-term, low-interest loans to the neediest African nations.

Initiatives by Saudi Arabia led to the establishment in 1976 of a US$1.2 billion International Fund for Agricultural Development (IFAD). This fund supports gifts and loans at low interest for long-term agricultural development in Asia, Africa and Latin America. It has been estimated that as a result of IFAD, agricultural output has increased by some 30 million tonnes per year – enough to feed an additional 200 million people. In addition, OPEC also sponsored a joint proposal in 1976 for setting up an annual fund of about US$800 million to assist Third World countries whose economies were hurt most severely by the five-fold increase in oil prices in the mid-1970s. Assistance from such a fund represents a price reduction in the cost of oil and the fund provides long-term interest-free loans to other developing countries to meet short-term balance of payments deficits and to pay for development projects and programmes. In general, the value of such loans and financial assistance is of inestimable value to poor countries.

Other regional development banks had been set up earlier to provide multilateral aid. The oldest and largest is the Inter-American Development ment Bank founded in 1959, which provides loans to Latin American countries at both market rates of interest and at subsidized rates depending on a combination of market financing and US government subsidy available. The AfDB, set up in 1964, has not been as successful in attracting capital as the Asian Development Bank (ADB) set up a year later. The ADB has obtained both market finance and government support, for example, from the US and Japan, to support its activities. A measure of its success is that the ADB now lends at market rates of interest to Asian nations. In Europe, two bodies provide multilateral aid – the Economic Development Fund and the European Investment Bank, both affiliated to the EU. These bodies extend loans and grants to other countries within the EU as well as to Greece and Turkey.

One final and increasingly significant aid agency is the group of institutions known collectively as non-governmental organizations (NGOs). While these may have humanitarian, religious and environmental objectives in their charter, NGOs have emerged as prominent channels of aid and development cooperation. Some examples of NGOs include Care, Community Aid Abroad, Oxfam and World Vision. These organizations obtain finances from governments as well as from private donations and fund-raising to support their development efforts. Official contributions to NGO activity have been rising and were about 5 per cent of ODA in 1987. The NGOs themselves provided about US$3.3 billion for development through private fund-raising in 1987. NGOs provide grass-roots support and their approach to development is consistent with the participatory principle. NGOs are active in health, welfare and education sectors. NGOs deliver aid professionally and have become a useful conduit between countries where political, economic and other ideologies of donors and recipients are at variance. NGOs have the advantage of being independent and are less likely to be instruments of purely national policy. Moreover, NGOs are less likely to fund projects that are unpromising and unlikely to succeed, nor will political pressures be brought to bear on NGO activity. Limited funds also mean that NGO projects are restricted in scope, thus eliminating grandiose prestige-driven projects in developing countries.

The activities in developed countries of civic minded people and humanitarians in organizing fund-raising activities such as 'Band Aid' and 'Live Aid' has brought the plight of developing countries into public consciousness. Such high-profile fund-raising activities not only raise important development funds but also demonstrate the global response to typically local problems. In this way too it is possible not only for governments to pledge matching funds but also for every lay person, young or old, rich or poor in developed countries, to do their bit for the development of economies and the alleviation of poverty in developing countries.

There are, of course, other alternatives to bilateral and multilateral financial assistance such as private investment schemes and tax concessions. While strictly such methods are not considered 'aid', the benefits and advantages are no less important. Private investments, while motivated by self-interest, also promote development and growth in developing countries. Such investments are usually accompanied by managerial skills, access to international markets and high levels of technical knowledge. For this reason, private investors seek rates of returns that are higher than average and usually choose their target countries with care. Some developing countries, conscious of the influence of foreign firms, can be reticent in allowing private investment. But where some compromise can be worked out, the results can be beneficial to both parties. Moreover, donor countries of the DAC would not have to draw upon public funds for such activities which means that aid funds can be used more profitably in other projects.

Alternatives which involve some degree of government participation include:

- revising tariffs in developed countries so that commodities and manufactured products from LDCs can be given preference in the developed world;
- establishing schemes to stabilize commodity prices of primary products and guaranteeing prices through price support and buffer stocks;
- guaranteeing bond issues of underdeveloped countries in international capital markets; and,
- using the so-called 'new money' created by the International Monetary Fund (IMF) known as 'Special Drawing Rights' (SDRs) to supplement the international reserves of gold and foreign currency of member countries.

Aid, development experiences and the debt crisis

While foreign aid has historical antecedents, the modern counterparts of such activities are of particular interest here. Of relevance is the period just after the Second World War when many countries throughout the world were either rebuilding their war-torn economies or emerging economies were struggling to overcome domestic difficulties and shortages resulting from world-wide economic depression. It is from the experiences of this and later periods that many of the examples of the effectiveness or otherwise of aid can be discerned. Although the impact of aid may have different effects on different countries due to varied backgrounds and 'degrees' of need, some common elements may be observed.

Economists have disagreed among themselves as to the impact of foreign aid on less developed countries. The traditional view is that aid has helped and promoted growth and has brought about the structural transformation of the economies of many developing countries. This view is especially tenable where the countries being assisted have either an economy that has been totally wrecked by warfare and political conflict thereby leaving a void, or where there has been a complete change from one economic system to another. Examples of the former are few in that, whatever the degree of conflict, there still remains some underlying structure on which the economy rests. In the latter case, the change-over from a market-based to a centralized socialist state provides a good example. So the traditional view that aid has brought about a structural transformation of the economy has to be accepted conditionally. While the aid–growth thesis is highly plausible, it is difficult to prove or support with evidence.

An opposing view to the first hypothesis is that aid does not encourage faster growth but rather retards growth. The argument for this rests on the view that as foreign resources are introduced into an economy, domestic savings and investments are prevented from participating in the growing economy. Additionally, the developing country is saddled with a growing balance of payments problem since it has to repay foreign aid in the form of loans and foreign 'tied' aid in the form of goods and services which have to be purchased from donor countries. This 'two-prong' onslaught arising from foreign aid results in larger balance of payment deficits and rapidly dwindling foreign exchange resources which could be better employed elsewhere in the developing economy.

At a local level, foreign aid has been blamed for exacerbating the differentials in two sectors of the developing economy. The focus on the urban sector of the economy and stimulating its growth has left the rural sector vulnerable. There is no stimulus for growth in the rural sector because the attraction to the modern urban sector appears more profitable. Heavily capitalized, profitable ventures and high wages in the urban sector also attract foreign investment and further aid in the form of long-term loans and grants. The rural sector, on the other hand, is unable to compete since it cannot provide such attractive rates of return as the urban sector. Inevitably there will be increasing gaps in incomes and standards of living not only between the urban–rural sectors but also within the urban economy itself – where the gap between the rich and poor will tend to widen. This increasing income inequality and the reduction in savings can be traced directly to foreign aid. Moreover, rather than assisting in filling gaps and relieving economic bottlenecks, the result has the opposite effect since foreign resources are now used to support the economy. New gaps are produced within the economy, for example, rural–urban differences and emergent formal–informal sectors in the urban economy. The argument leads to the conclusion that aid can have a negative effect and may be anti-developmental.

On the donor side of the equation, it may be said that after three or four decades of aid and foreign assistance, aid weariness and aid fatigue may now be setting in. The view is not so much that the donors have lost their generosity but rather domestic issues in donor countries themselves have led to a reduction in aid giving. The economic climates of the 1970s and late 1980s have been such that donor countries themselves are struggling to rationalize the *raison d'être* for foreign aid. Domestic issues such as inflation, unemployment and balance of payments problems weigh heavily in donor economies, problems which, paradoxically, recipient countries are also striving to solve and overcome.

Part of the solution to this problem of the shortage of financial resources is to look towards those newly emerging economies, especially those which have accumulated a largesse from petroleum production, where new aid relations are being forged. The IFAD, for example, has set aside up to US$1 billion to finance balance of payments adjustments for needy developing countries. The lower total value of aid from developed countries has also been accompanied by a change in emphasis in the use of aid resources. No longer is the focus on industrial growth, but rather the emphasis is on basic needs of food, shelter and mobility and on rural development. To lessen the impact of particular donors determining the direction of aid, foreign aid should be channelled through multilateral assistance agencies and through NGOs. The bad side-effects of 'tied' aid will be lessened considerably. It seems that to achieve the greatest impact in development assistance, foreign aid should be in the form of outright grants and concessional loans rather than in any other form. Grants and loans will afford more autonomy in the use of these resources since these may be redirected to those sectors of the developing economy most in need. A final suggestion is the view that donor countries can assist much more simply by reducing the tariff and non-tariff trade barriers that exist and which have been used against developing countries. Such reductions in trade barriers can have a dynamic effect in boosting trade.

Aid to developing countries leaves many countries indebted in more ways than one. Leaving aside the moral issues such as obligations and patronage, the economic imperatives of external debt are that most developing economies have to repay aid that came in the form of loans. One way to appreciate the magnitude of this problem is to examine some World Bank data. According to the World Bank, debt includes private non-guaranteed debts. Public loans are external obligations which national governments and their agencies and autonomous public agencies have borrowed. It is to be noted that public guaranteed loans are external obligations of private debtors that are guaranteed by national governments, while private non-guaranteed loans are external obligations of private debtors that are not guaranteed for repayment by a public entity.

The abrupt end of abundant flows of financial resources to developing countries in 1982 set off a debt crisis. Increased private flows went to meet debt-servicing needs of debtor countries leaving little additional capital for investment and sustained growth. The persistence of the crisis throughout the 1980s meant lower investments, lower growth and higher inflation rates in many developing countries. Contributing to this were the large fiscal deficits, over-valuation of currencies and a bias against exports in developing economies. A current account deficit refers to the excess of import payments over export receipts for goods and services received, while a capital account deficit refers to the excess receipts of foreign private and public lending and investment in the repayment of the principal and interest on former loans and investments.

Externally, the rapid increase in interest rates around the world, falling commodity prices and a world recession did not help the cause of developing countries. Table 3.3 gives indicators of external debt for developing economies for 1970–89. These data show that net transfers to low- and middle-income countries were negative in the second half of the 1980s. Interest payments as a share of total export receipts in the 1983–9 period in most cases have more than doubled compared with those at the beginning of the 1970–5 period. This is not surprising given that the total external debt as a share of the GNP has been doubling every six years in the periods from 1970–5, 1976–82 and 1983–9.

Table 3.3 Indicators of external debt for developing economies, 1970–89 (average percentage for period)

	Total external debt[1]			Interest payments[2]			Net transfers		
	1970–5	1976–82	1983–9	1970–5	1976–82	1983–9	1970–5	1976–82	1983–9
Low-income	10.2	14.2	28.5	2.9	4.3	9.8	1.1	1.2	0.7
Low income (excl. China and India)	20.5	28.5	60.7	2.9	5.3	11.8	2.7	2.4	1.0
Middle-income	18.6	34.6	54.9	5.1	11.0	15.4	1.9	1.9	-2.7
Argentina	20.1	46.1	80.3	14.1	17.9	41.6	-0.3	2.7	-5.4
Brazil	16.3	28.2	42.0	12.1	28.5	30.3	3.3	0.8	-2.5
Morocco	18.6	55.1	109.5	2.8	13.0	17.1	1.8	6.8	-1.7
Philippines	20.7	45.8	79.2	4.2	14.1	20.5	1.2	1.8	-3.4

Source: World Bank (1991) *World Development Report 1991*, New York: Oxford University Press, Table 6.2, p. 125
Notes: Values are yearly averages calculated for the period; economy averages are weighted using the share of GNP in 1981.
  [1] As a share of GNP.
  [2] As a share of total export receipts.

Most low-income countries owe debts to official creditors, bilateral and multilateral agencies, with a large part of the private export credit being officially guaranteed. Official creditors have engaged in debt forgiveness and rescheduling and provided additional new flows at highly concessional rates. Bilateral creditors have rescheduled loan repayments under the so-called Toronto terms by offering highly concessional rates. Under these terms, bilateral official creditors who have extended non-concessional loans may choose between cancelling one-third of the amount, adopting a longer repayment period as used for concessional debt (25 years' maturity and a grace period of 14 years) or cutting the interest rate. For commercial debts, under the Brady Initiative official creditors have offered to support debt and reduce the debt servicing for countries that have adopted adjustment programmes and who have taken measures to encourage direct foreign investments and the repatriation of capital. By 1990, new debt agreements based on the Brady Initiative had been implemented in Costa Rica, the Philippines and Mexico (see Case study C), and negotiations were under way in Morocco, Uruguay and Venezuela. In the case of Mexico, for example, since the announcement of the agreement in July 1989, real interest rates have declined substantially and capital inflows have risen.

The debt problem associated with aid has also prompted a re-evaluation of what development assistance can or cannot do. There seems to have been an over-optimistic notion of what foreign aid could achieve. In the context of the aid programme in the period after the Second World War, the tangible effects of military aid were quite evident with many recipient countries quickly building up their military capabilities both in the democratic and the eastern bloc countries. But the assistance directed towards economic development has been less spectacular given that the results take a longer time and require also a transformation of society. This slow progress towards rapid economic development has also given rise to scepticism as to the merits of foreign aid. For donor countries, it seems that at times foreign aid can be politically unrewarding. This is because the recipient countries are invariably those that have just attained their independence and having come with a colonial heritage these recipient countries may wish to preserve their newly found independence by pursuing policies which they perceive to be in their own interests. Such interests may conflict with those of donor countries. The donors, in using aid as an instrument

of foreign policy, may be reluctant to give financial or other assistance
from which no tangible political gain is readily apparent.

Case study C

The 1990 Mexico Debt Agreement

Since the 1982 debt crisis, Mexico has negotiated rescheduling of
loan repayments and organized new money packages in 1983–4
and in 1986–7. These agreements, however, failed to provide
medium-term relief and debt servicing of external lenders and
creditors. In 1985, Mexico introduced important reforms of external trade and the financial sector, privatized many state-owned
enterprises and overhauled regulations on direct foreign investment. Despite these efforts, external debt continued to increase.
Large external transfers created uncertainty about future exchange
rates and tax policies. To prevent capital flight, Mexico paid very
high interest rates in its domestic debt, thus threatening the results
of fiscal reform that had been undertaken in recent years.

In March 1990, Mexico and its commercial creditors implemented a debt restructuring agreement. Banks could choose from a
number of options that included 'new money' and two facilities
for reducing debt and debt service – an exchange of discount
bonds against outstanding debt, or an exchange of bonds against
outstanding debt without any discount but bearing a fixed interest
rate. About 13 per cent of creditors chose the new-money option,
40 per cent chose the discount bond and 47 per cent chose the part
bond at 6.25 per cent interest. Collateral funds were drawn from
country reserves and loans from the IMF, World Bank and Japan.
Participating banks were ineligible to take part in a new debt–
equity swap programme linked to the privatization of public
enterprises.

The debt restructuring agreement was expected to reduce
Mexico's net transfers abroad by about US$4 billion a year during
the period 1989–94. About half the reduction comes from the
rescheduling of amortization. These reductions will improve Mexico's fiscal position and should have a beneficial effect on growth.
The agreement has also altered expectations by diminishing the

Case study C (*continued*)

> uncertainty about future exchange rate and tax policies. In addition, real interest rates have declined substantially and capital inflows revived.

It has also been felt that the quality and delivery of aid can be significant in shaping development. The experience internationally indicates that the quantity and quality of aid go hand in hand. Donor countries who give more also seem to have the best quality aid programmes. Conversely, those who give least seem to have more self-serving and poorer quality programmes. The quality of the official aid programme is governed both by the amount and objectives of such aid. The central aim of aid programmes must be developmental – to assist in the social and development of LDCs in the world. Serving commercial and foreign policy interests, while an integral part of the objective of aid giving, must take on a secondary role to the main objective.

As an example, many official statements by governments about the aid programme highlight three objectives: humanitarian and developmental, commercial, and political, with the implicit assumption that the aid programme can and should pursue all three simultaneously. The reality, however, is that this is an impractical objective because each of these leads to different sets of priorities in terms of the sectors and types of projects that need active support. Thus, a humanitarian or development objective would lead, for example, to the Horn of Africa being high on the list of priority areas whereas under the other two objectives it may not feature at all. This is because to the developed world there are little or no commercial and political interests in the Horn of Africa. There is thus a need to decide on the priority to be given to each of the three objectives noted above.

The quality of aid received also hinges on its delivery and the results that have been achieved. In general, the efficiency and effectiveness of bilateral country programmes have progressed well because donor and recipient countries have worked closely in targeting the aid effort. Where such aid reaches the people most in need and produces results in project completion, aid is said to be effective. Also where the quantum of aid given has remained steady and reaches the target areas or sectors relatively undiminished, aid delivery is said to be efficient.

For these circumstances to prevail donor countries usually have a good knowledge of the areas and sectors to target. Where there is pressure to divert such aid for short-term political and commercial interests, the aid programme may be weakened and ultimately become fragmented thus lowering the quality of aid. Some donor countries contribute more to multilateral agencies. Here there is less control in targeting aid although similar comments apply with respect to the effectiveness and efficiency of aid delivery. It seems that all aid should focus on combating poverty, prompting a shift in the sectoral focus of aid to the most needy countries and to sectors in which there is an emphasis on community participation. In these instances, where aid addresses basic needs of food, shelter and movement, and where programmes are designed to help define local priorities and to strengthen their human rights issues, then it may be said that the aid programmes have been designed effectively and are being delivered efficiently. In this way, multilateral agencies can and have been able to enhance the quality of foreign assistance.

Figure 3.2 The many faces of the aid question
Source: *Development Forum*, vol. 3, no. 2: 11

Despite all the efforts at development assistance, the battle against ignorance must be fought not only in the developing world but also within donor countries. Figure 3.2 is a poster produced by War on Want in the UK where the message is graphically brought home to all concerned.

Aid politics and aid as imperialism

Any summary of the donors and recipients would be incomplete without a discussion of the objectives and motivations of the donors and the perception of recipients in accepting assistance. The political, strategic and self-interested intentions of donors have to be matched against the economic, moral and political benefits of beneficiaries. In 1971, Theresa Hayter wrote a book entitled *Aid as Imperialism* in which she said that aid can only be explained in terms of an attempt to preserve the capitalist system in the Third World. This is a 'dependency' viewpoint in which the state is simply functional to capitalism and where the 'local classes' in developing countries become wholly symbiotic to foreign capital. Such a view therefore suggests the need to examine both sides of the aid coin – why donors give and why recipients accept aid. Understanding these different motives can help explain when aid is effective and when it is not.

Donors give aid for political and strategic reasons as well as for economic self-interest reasons. Aid giving is also motivated by moral and humanitarian reasons especially for emergency relief, famine and civil unrest. Many donor nations assist others without expecting corresponding benefits in return, whether political, strategic or economic. Such altruistic objectives are neither easily defined nor assessed but the spread of goodwill and assisting development to reduce the possibility in the long term of serious political unrest and enhance the stability of governments have indirect benefits all round. The history of aid giving has demonstrated that not all aid is profoundly altruistic. The reconstruction of Europe after the Second World War through the Marshall Plan was ostensibly to combat communism, as has been the assistance to South Korea, South Vietnam, to countries in Latin America, the Middle East and eastern Europe. The shifting theatres of conflict reflect the changing political and strategic interests of the major players led primarily by the US. As the assistance furthers the interests of the donor country, the flow of funds varies according to the donor's assessment of the political situation and *not* the relative needs of the

The Million Dollar Man

*Yes, I am the expert
As you can well see
Arriving by jet plane
With my HYV**

*It's straight to the hotel
Decisions today
I'm meeting those chaps
From the M o A**

*Let's grow carnations
— now wouldn't that be
A nice way of raising
Their low GNP**

*Then off to the village
(I hope it's not far)
First on with the shorts
And my SLR**

*I can't speak the lingo
Phew, what a hot day
— But I saw a nice farmer
And things seem OK*

*So I write my report
And I pocket my fee
Then it's time to depart
For the next LDC**

The Village

*Now who was that man?
And why did he come?
Stayed just half an hour
Only met the chief's son*

The Government

*What's this he's left us?
Four volumes or more!
We'll need a new expert
To say what it's for*

KEY TO THE EXPERT

HYV	High Yielding Variety
M o A	Ministry of Agriculture
GNP	Gross National Product
SLR	Single lens reflex (camera)
LDC	Less Developed country

Figure 3.3 The delivery of aid and the expert: a cynic's view

Source: M. Carr (ed.) (1985) *The AT Reader. Theory and Practice in Appropriate Technology*, New York: Intermediate Technology Development Corporation of America: 339

receiving country. Figure 3.3 gives one cynic's view of the delivery of aid and the expert.

In economic terms the reasons for giving aid is more easily discussed. External sources of loans and grants can play a critical role in supplementing domestic resources to provide developing countries with the means to develop. Such sources reduce the need to use up limited foreign exchange to finance the import of capital and intermediate goods. Also there is invariably a shortage of domestic savings with the result that domestic investment opportunities have to be foregone in favour of other more pressing needs such as poverty and food shortages. With no savings and no investments, there can be no growth in the economy. When loans are tied to particular projects, developing countries acquire both modern technology and personnel who are then able to pass on technical knowledge and skills to those countries. Such transfers of technological skills are invaluable in enlarging the technology and skills base of developing countries. Foreign aid in the form of financial assistance, technical assistance or food aid releases resources which are otherwise unavailable locally. Economists have identified the 'absorptive capacity' of a country as its ability to use funds wisely and productively. Unfortunately donors prescribe how these funds are to be used and this decision is based on the donor's overall economic assessment of the country. The economic motives are that definite benefits accrue to donor countries as a result of the aid programme because more donor countries are giving loans instead of outright grants and a high proportion of aid is 'tied' to the exports of donor countries. Most aid funds therefore are retained in donor countries by paying for export sales, enriching its citizen consultants and aiding its bureaucracy (see Case study D). Apart from long-term returns on capital, donor countries also have the additional objectives of securing a supply of raw materials, 'cornering' a particular market, thus putting themselves in a stronger position *vis-à-vis* their competitors.

Case study D

Boomerang aid from Australia

In a publication circulated by the Australian Minister for Foreign Affairs and Trade, Senator Gareth Evans, entitled *Australia's Aid Program: Five Years after Jackson* (1991), it was observed that:

> Our overseas aid expenditure is both altruistic and in our own interests, and is capable of being looked at from both these perspectives. All Australia's aid, to qualify for that description under international accounting rules, has to be altruistic: alleviating poverty and distress, promoting development or both. But equally, all Australian aid can be seen as promoting one or more very direct and very real Australian interests. . . . What is important is that these altruistic and self-interested perspectives complement each other.

The synergy between altruism and self-interest can be very powerful and leads to 'win–win' situations where benefits flow to both donors and recipients. Australian agricultural output is in many ways complementary to those food needs of the Third World, for example, rice and other foodstuffs. Data on trade and aid show that the latter is a 'trade promoting boomerang' in that it promotes broad-based economic growth in less developed countries (LDCs) which also encourages the growth of imports by these countries. Finally, Australian aid is an investment in the future, especially that aid which meets humanitarian goals – for example, growth through poverty targeted aid which may be beneficial to Australia in the future.

The Australian record in development assistance is highly creditable. Australian aid has risen at an increasing rate and matches the average given by members of the DAC. Following the end of the Second World War the Australian commitment to foreign aid was less than US$9.8 million annually, after the Korean War it doubled and then more than trebled to US$37.5 million by 1959. By 1965 it had doubled again to US$75 million and has been doubling at approximately six-year intervals ever since to reach about US$600 million by 1983. If the trend continues, ODA

Case study D (*continued*)

will reach US$1.5 billion by the middle of the 1990s. However, ODA as a percentage of GNP has steadily declined from 0.53 per cent during the early 1970s through to 0.33 per cent by mid-1980 with the projection that this will fall even further in the late 1990s. Thus, the UN target ODA of 0.70 per cent of GNP will be unattainable in the foreseeable future in Australia's case. This is because of the dominance of Papua New Guinea in Australia's aid budget, Australia's growing contribution to multilateral agencies and Australia's economic difficulties in the 1990s.

An estimate by AIDAB (renamed AusAID in 1995) – the Australian International Development Assistance Bureau (or Australian Agency for International Development) – shows that the aid programme as a whole generated purchases of Australian goods and services to the value of 87 per cent of total aid expenditure. In other words, 87 per cent of the aid given came back to Australia. The aid programme by promoting growth and stability in developing countries also promoted the expansion of important new markets for Australia, aid generated the purchase of Australian goods and services. These observations are given further support in the distribution of Australian aid which is concentrated in countries in the Asian region where commercial and foreign policy interests are best served.

Australian policy seems to be that aid should promote growth which should lead to trade. This policy is given substance through the Development Import Finance Facility (DIFF) which provides subsidies to Australian suppliers of capital goods and gives Australian firms the competitive edge in bidding for contracts internationally. Detractors of DIFF, however, see this as distorting aid priorities. Between 1980 and 1986 DIFF was a small component of the aid budget (around US$12 million), but by 1991–2 it had grown to US$69.8 million, about 7 per cent of the total aid programme, and accounted for US$90 million in the aid budget in 1992–3. DIFF, however, is 'bad' aid because its basic motivation is commercial not developmental. There is a bias towards countries like China, Indonesia, India and Thailand where 84 per cent of DIFF subsidies have been directed. The support for the purchase of capital goods and towards infrastructural projects means that the economic trickle-down effects do not reach the poor. DIFF also disadvantages Third World

Case study D (*continued*)

> manufacturers whose governments are unable to support them in the same way as the Australian government. The concentration on capital goods also means that developing countries are obliged to borrow to finance purchases thus contributing to further indebtedness. The argument in favour of DIFF is that it enables scarce development funds to be spread out further than would otherwise be possible in the usual manner of providing aid.

The reasons why countries accept aid will be discussed to a lesser extent mainly because the motives and interests here can be seen to conflict directly with donor countries. In economic terms developing countries accept uncritically that economic assistance is important if they are to develop economically. The importance stems from the fact that aid will supplement scarce resources, help to transform the structure of the economy and contribute towards a self-sustaining growing economy. However, because a greater proportion of the aid is in the form of loans, these developing countries are saddled with a debt problem of huge proportions. The preference is for outright grants, in larger amounts, and low-cost loans with little or no conditions attached. This will mean the abolition of 'tied' aid and the granting of autonomy in the use of such aid for developing countries. Herein lies the conflict of interest between donor and recipient countries – a problem with no immediate solution. The trend towards a hardening of aid conditions given the rising interest rates and shorter repayment periods makes the acceptance of foreign aid less attractive. Aid is also seen in some countries as providing some form of political leverage in helping prop up 'regimes' so that the status quo in terms of trading and other economic relations can proceed uninterrupted. The massive injection of funds for military and security assistance in South Korea and South Vietnam were examples of this aspect in the aid programme.

Finally, developing countries consider that rich nations have an 'obligation' to support the economic and social development of the Third World because of past colonial exploitation. At the same time, developing countries yearn for the freedom to be able to allocate and use aid funds as they wish. The threat of defaulting on loans and/or nationalization of industries are potential weapons developing countries have considered. A pragmatic response has, however, prevented defaulting or nationalization as the consequences for the developing nations would be too serious to contemplate, for example, ostracism by the international community or self-imposed isolation. It is more likely that defaulting on loans will arise from the inability to generate sufficient income to service them. In any case, developing countries cannot be considered to be receiving resources when the repayment on loans has to depend on the flow of aid.

The contribution of aid to development can now be summarized. There are circumstances when aid can be ineffective and when it is most effective. Taking the negative argument first, aid can be ineffective in some of the following circumstances:

- Where aid permits countries to postpone improving macro-economic management, for example, restructuring their economies and fiscal reform. Tanzania's experiments with disastrous rural programmes were largely funded by foreign aid. Pakistan avoided fiscal reform for political reasons.
- Aid at times replaces domestic savings and fiscal discipline thereby producing some measure of aid dependency. If aid is allowed to depress agricultural prices and to postpone investments in rural infrastructure it can lead to long-term problems.
- The uncertainty in the financial flow of aid as a response to political and economic imperatives of bilateral funding can be destabilizing and interrupts development programmes as has happened in Egypt, India and Pakistan.
- Tied aid leads to the transfer of inappropriate technology and skills which substantially reduces actual resource transfer to developing countries.
- Changes in policy advice from funding agencies can lead to conflicts in project implementation so that programmes now in place as a result of previous policies would need to be replaced by new policies and programmes. Such changes add to the cost of aid and retards progress.
- Foreign aid is ineffective where it is redirected towards the repayment of loans, correcting problems with the balance of payments and redressing deficits in other sectors of the economy rather than on development projects themselves. Argentina, Uruguay, Chile, Algeria and Indonesia provide examples.

On the other hand, aid is effective in the following circumstances:

- Aid promotes the development of countries that are at least partially modernized, for example, Greece, Israel, Spain and the former Yugoslavia.
- Aid provides external resources for investment and finances for projects which would otherwise not be undertaken. The massive and sustained flow of financial aid has led to successful development in Jordan, South Korea and Taiwan. On the other hand, discussions and proposals for aid from developing countries inform developed countries about reforms taking place. This improves the access that developing countries have to capital and direct foreign investment as in the case of South Korea, Malaysia and Thailand.
- Socialist countries employing draconian measures such as nationalization, state investment and rationed consumption have brought about some measure of development through an injection of foreign aid. China, North Korea and Vietnam are examples.
- In economies whose policies encourage exports it has been shown that a modest infusion of foreign aid can bring about economic progress, for example, Mexico, Pakistan and Thailand. Aid improves economic reform, for example, in Chile, South Korea and Turkey. Thus, domestic policies, institutions and administrative capacity should be supportive of aid projects for ultimate success to be achieved.
- Project assistance helps expand infrastructure – roads, railways, ports and power-generating facilities. There is also a transfer of skills when massive infrastructural and industrial projects are undertaken as in Colombia, South Korea, Mexico, Pakistan and Thailand. Bangladesh's Grameen Bank, which provides loans to small family investors, is an example of the successful use of economic knowledge at a local scale.
- Aid which is supportive of economic, social and environmental policies can result in successful programmes, for example, the reduction of poverty in Bolivia, Côte d'Ivoire and Malaysia, or achievements in education and health as in Pakistan.

Conclusion

The aid question has often been coupled with the issue of trade. Many experts take the view that opportunities for profitable trading with developed countries are more valuable to Third World economic growth than any quantitative increases in development assistance. Such a view appears to be a pipedream which would not bear close scrutiny since a move in this direction would be against the economic self-interests of developed countries. However, on closer examination it seems that the liberalizing of trade as a form of aid and the restructuring of trade and aid policies may go a long way in assisting economic development. The realities of global interdependence in the new century would suggest that there will be long-term economic, political and social benefits to be achieved from the rapid economic development of the Third World. To appreciate these effects there is a need to examine the growing global interdependence in the form of technology transfer, appropriate technology and countertrade, themes taken up in the next chapter.

Key ideas

1 Aid as the transfer of resources takes many forms including grants, loans and technical services. Such aid is necessary in order for developing countries to attain a stage of sustained economic growth and development.
2 The pattern of ODA to developing countries fell in the decade of the 1980s. ODA as a share of donor GNP of OECD countries in 1990 was about 0.36 per cent.
3 Aid has been delivered by official agencies of the UN and the World Bank. However, increasingly aid is being delivered by NGOs as well as through Third World agencies.
4 The impact and effectiveness of aid varies depending on the recipient country, its socio-economic and political background and degrees of need.
5 The reasons countries give aid can be traced to political, strategic and self-interested motives. Recipients of aid, however, accept aid because of the perceived economic, moral and political benefits. Donor and recipient motivations may sometimes be entirely at variance in so far as aid is concerned.
6 Aid in the form of loans presents developing countries with a continuing debt problem. The preference is for aid in the form of outright grants.

4
Countertrade and technology transfer

Introduction

One of the unintended consequences of international aid is the growing indebtedness among Third World countries. Such indebtedness has arisen from the very nature of aid programmes where developing countries are forced to pursue certain courses of action even though such actions might eventually lead to further debt and loss of independence. Moreover, the already weak economies might be further burdened by poor terms of trade and balance of payments problems. To overcome the debt crisis and in an attempt to pay for imported goods and services many countries, especially those in the eastern bloc, have engaged in alternate forms of trade known as countertrade or counterpurchase and bartering goods. The country in debt may enter into an agreement with another country for either payment in kind (exchange goods) or to offset debts by offering generous investment incentives.

This chapter considers the problems and opportunities presented by countertrade, described by some as an inefficient and costly form of protectionist dominated trade. Apart from trade, this chapter will also consider the view that there has been a shift in emphasis from mere foreign investment to that of outright technology transfer. This is because in a world of unequal bargaining positions, the technology from the developed world may be required in the developing world if the resources of the latter are to be put to more efficient use. The end result would seem to be a 'win–win' situation for both participants. In the first section a definition of countertrade is presented and the reasons for it being considered an egregious form of trade are discussed. Then studies of five countries practising various forms of countertrade are considered together with an evaluation of the prospects of countertrade. In the second section the transfer of technology to developing countries is evaluated in terms of the appropriateness of such technology. The need for technology transfer is discussed as well as its so-called 'back-wash effects'. This is followed by a discussion of the costs of technology transfer including strategies and problems as well as the role of multi-national corporations (MNCs) in technology transfer. In the third section 'relationship enterprises' as an alternative model to the MNC are described. There is also an examination of Asian MNCs, including their role in the development of intermediate economies.

Countertrade

The era of a 'new mercantilism' has resulted in many less developed countries (LDCs) striving for a share of a diminishing and increasingly protected international trade. In addition, the lack of 'hard' currency exacerbated by the growing indebtedness of many LDCs has meant that these countries are precluded, and in some cases restricted, from trading in the open international market. The mechanism of 'countertrade' – a means of obtaining goods and services without necessarily using money – has opened up opportunities for these LDCs to be involved in the global economy. Many LDCs are in debt and a healthy export perfor-mance is one means of paying off foreign debt. Poor trade performance is a major obstacle to economic growth. Given this economic environ-ment, a 'new mercantilism' has emerged whereby nations are trying to gain advantages for themselves by selling more than they have bought in order to build up a favourable trade balance. As a result, ingenious schemes outside traditional mainstream trade practices have emerged and these include new forms of barter and countertrade. Such interna-tional trade is one where there is a conditional linkage between exports and imports: one country will buy from another if the latter country is prepared to buy something else in return.

Definition

Countertrade may be described as a means of obtaining goods or services without necessarily using money. Countertrade consists of a set of parallel obligations where each party undertakes to sell goods or services to each other in separate but related transactions. This reciprocity is the result of the mutual needs of both the foreign exporter and local buyer. Here we refer to the foreign exporter as the party from a developed country whereas the local buyer refers to the importer in the LDC. Countertrade is sometimes seen as a form of currency exchange, to collect a debt, or a means of finance where foreign currencies do not change hands. Goods are exchanged between two, three or more parties until everyone is satisfied with the outcome.

In general, all forms of countertrade share the common characteristic of a foreign exporter selling to a local buyer in one contract. In a separate but related agreement the foreign exporter enters a contract to purchase from the local buyer goods or services equal to an agreed percentage value of the original sales contract. A typology of various forms of countertrade is shown in Figure 4.1.

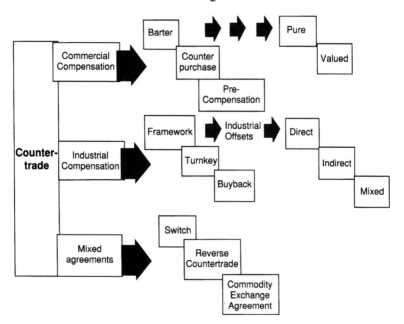

Figure 4.1 A typology of countertrade

Barter transactions involve the direct exchange of goods or services between the foreign exporter and local buyer and normally does not involve the exchange of money or the use of third parties. But with the creative use of financial mechanisms and settlement accounting procedures, pure or true barter of goods is no longer common. Pure or true barter is a simple exchange of goods whereas valued barter refers to some value put on the exchanged goods. The latter is by far the more common of the two transactions. However, the simultaneous double coincidence of needs between the parties is difficult to achieve, thus limiting the usefulness of this form of transaction in the modern business world. Barter arrangements among suppliers of homogeneous goods (for example, oil, mineral ores) to save transport costs are called swaps.

Counterpurchase is the most common form of countertrade and is where the foreign exporter provides goods in one contract and undertakes to purchase goods for export from a local buyer in a second contract. The two contracts are independent of each other as are the payments. The total time involved is between one and five years and the value of counter-delivered goods is usually less than 100 per cent of the value of the original sales contract. Each party pays cash for goods or services received. Often, a list of goods is identified to be selected for purchase.

Compensation is a generic term that denotes partial exchange. A foreign investor receives payment either in full or in part from the local buyer in the form of goods in exchange. Thus, buyback involves a foreign investor providing industrial machinery and as a direct result is committed to the purchase of goods produced by these machines. Sometimes offtake and industrial cooperation are terms used to describe these agreements. For example, the sale of a 'turnkey' oil refinery is paid for by the resultant refined oil. 'Turnkey' refers to the fact that a foreign investor will build the facility to completion and hand over the key to the local partner to operate. The agreement usually requires the foreign investor to purchase an amount of the resultant product that exceeds the value of the original contract over a period of five to ten years. Most buyback terms contain a most favoured customer (MFC) clause where the local buyer will grant the foreign investor the most favoured price for the same goods when compared to another purchaser. Other terms in the agreement may include:

- co-production in the local buyer's country based on bilateral agreements;
- licensed production;

- subcontractor production of parts or components that go into the foreign investor's product;
- capital investments to expand joint ventures in the buyer country; and
- transfer of technology with research and development (R&D) in the buyer country.

Mixed agreements include transactions known as switch trading or clearings. This consists of bilateral trade-and-payment agreements. In bilateral trade the value of goods exported from country A to B is not actually paid for but is credited to A on a clearing account and vice versa as goods are shipped in the other direction. Usually outstanding balance and interest payment are cleared or settled at the end of each accounting year. Where the goods have a readily accessible commodity market (for example, oil) the exporter may engage a dealer to market the goods. As a string of dealers may be involved, property in this countertraded good is passed by the transfer of negotiable bills of lading, the document containing the terms of a contract of carriage between the consignor of the cargo and the owner of the ship transporting the goods.

The picture is, however, more complicated where no commodity markets exist for the countertraded goods. For example, the foreign exporter may be unable to obtain payment from the local buyer because the country will not permit the transfer price for lack of hard currency or because of 'blocked funds'. However, if the local buyer's country has a trade surplus with a third country on a bilateral clearing agreement, this credit surplus may be used by way of a switch transaction for payment due to the foreign exporter.

A switch transaction occurs after counter deliveries of the products begin. Because of difficulties with commodities having no ready markets, discounts of up to 40 per cent are given to trading houses or switch dealers. These trading companies maintain their own private networks that offer a ready market for discounted countertraded products. Switch trading and barter have been used as legal methods to avoid currency controls and take advantage of bilateral trade arrangements between countries.

Country studies

There are common reasons why countries engage in countertrade. This trend towards countertrade is spurred on not only by the rise of protectionism caused by a decline in confidence in the world trading system but also by a growing debt crisis among LDCs. Indebtedness inhibits exports, contributes to a decline in foreign exchange earnings which results in an inability to service foreign debt commitments. This apart, countries need to maintain existing market shares as well as penetrate new markets. While many LDCs desire sustainable growth in their development programmes, the accumulation of 'hard' currency is a way to foster growth in LDCs as much as to reduce trade deficits. By this means the economy will be able to maintain or even increase employment, reduce the technology gap and stabilize prices of export commodities. For the developed world, however, enlarging countertrade may mean the maintenance of sales volumes, market shares and profit levels in international trade. Also to prevent LDCs defaulting on their debts, developed countries may assist LDCs trade their way out of their financial problems.

The experience of practitioners in Asia and eastern Europe has suggested that countertrade may be a useful market development tactic rather than a problem to be avoided. Countertrade may be the answer to both cheaper imports and increased exports. However, it is unclear whether it is countertrade itself that is leading to an increase in business among developing countries. Rather it seems that economic conditions prevailing in many parts of the world may encourage growth in South–South trade. Countertrade happens to be one of the mechanisms used to foster the growth of international trade. Countertrade has offered a major opportunity to enter new markets and to consolidate positions in old markets. Indeed, countertrade adds to the total volume of trade in terms either of value-added or new products for exports and net growth of existing export goods.

However, it may be said that countertrade puts off the need for LDCs to correct the basic causes of mediocre to poor export performance and may delay internal micro- and macro-economic reforms. Also countertrade is hardly an innovation that will expand trade between developed and developing countries or make it more dynamic. In a sense countertrade is backward in its return to bilateralism through a modern form of a barter system and restricting choice in a way that is very much reminiscent of war-time economies and compartmentalized markets. Countertrade may be regarded as a relatively inefficient form of trade and may be contrary to principles embodied in the General Agreement on Tariffs and Trade (GATT). Also the restrictive clauses and conditions in countertrade contracts do not give LDCs a better control of their foreign trade. Exporters who are given compensation agreements may refuse to accept them rather than enter contracts they are unable to fulfil.

More critically countertrade may work against attempts to modernize the economies of LDCs. It is a regression to bilateralism and distorts the comparative advantage of countries that practise it, but it also sustains and tends to perpetuate inefficient and uncompetitive industries in those countries that would otherwise perish. In enforcing a countertrade deal, there is the likelihood that the local buyer may quote unrealistically high prices for local goods, offer poor quality goods or foist on to the foreign exporter goods that are difficult to sell in world markets.

It is suggested that countertrade is likely to be a longer-term phenomenon than initially expected and will persist over the next 15–20 years. Variations in defining countertrade and the scarcity of published data make it difficult to estimate the volume of international commerce under countertrade. The International Monetary Fund (IMF) and GATT estimate that only 5 per cent of world trade comes within countertrade arrangements whereas the Organization for Economic Cooperation and Development (OECD) put it at 8 per cent of world trade. Other estimates claim that countertrade now touches between 20 and 30 per cent of all international trade with a projection that it could reach 50 per cent of all international commerce by the year 2000.

Whatever its commercial value, the upsurge in countertrade has not resulted from commercial advantage but rather as a pragmatic response of world commerce to the financial crisis of LDCs. Importing activities may be sustained only by guaranteeing the replacement of hard currency within a relatively short time. Moreover, most LDCs use countertrade as a temporary means of protectionism. Given the rather diffuse picture of the incidence, degree and pattern of countertrading activity, an examination of selected examples from South East Asia will highlight some practical problems.

Barter experiences in Myanmar

In Myanmar (formerly Burma), several state corporations are responsible for countertrade in specific sectors. By definition these are all state monopolies since Myanmar is a socialist republic. While there is no gazetted policy for countertrade, bilateral transactions, barter and counterpurchase are the types most frequently pursued. In three cases where barter failed – one for military aircraft, another for vehicle spares and another for milk powder – the disagio in the transaction left little margin and the foreign exporter had to withdraw. The disagio is a fee or commission which is an amount of money that has to be spent before the goods become saleable. This over-playing of the export–import ratio can rebound in a negative way on the foreign exporter. Fortunately, the transactions in these examples were only at a negotiations stage and no binding agreements had yet been entered into. Given the current state of the Myanmarese economy, it is expected that for the foreseeable future counterpurchase or barter in new transactions may increase in volume.

Thai countertrade

Countertrade in Thailand is used as a means to supplement normal trade activities. There is no law specifically on countertrade and as such each arrangement with a foreign exporter is treated on a case-by-case basis. The government does not interfere with countertrade deals made in the private sector. However, guidelines are given in terms of preferred exports (especially those suffering depressed prices and market access difficulties) and imports that may form part of the general government development planning activities. Data for 1986–7 show that Thai products like maize and rice have been exchanged for fertilizers, crude oil, gantry cranes and medical equipment on a valued countertrade basis. In one deal the Thai Ministry of Agriculture and Cooperatives (MOAC) failed to deliver a quantity of maize for Romanian fertilizer and had to terminate the contract even though Thailand had already bought 25 per cent of the contract volume. It appears that the problems which have arisen are economic rather than legal, for example, the pricing of non-standard goods for which no market reference price exists.

Malaysian countertrade

In Malaysia, countertrade is monitored by the government's *Unit Khas Countertrade*. While countertrade is strictly between private companies the government offers policy guidelines to firms concerning imports and exports. However, since 1983 the government has mandated counter-trade provisions for all government contracts. Countertrade arrange-ments are examined on a case-by-case basis. The emphasis is on those transactions that result in economic growth and development for the country. Sales of certain primary commodities, trade with countries experiencing persistent deficits and developing non-traditional markets provide the impetus for countertrade. Although modest in volume,

Malaysian deals reveal a wide variety of products and partners. Malaysia has countertraded oil, palm oil, textiles, wood and electrical goods for patrol boats from South Korea, tin and rubber to former Yugoslavia for circuit breakers and power lines, oil to Brazil for iron ore, and palm oil to Pakistan and Thailand in exchange for rice.

Singapore countertrade

Singapore has become the dominant centre for countertrade in Asia by actively seizing commercial opportunities arising out of the promotion of such trade in South and South East Asia. The Singapore government has provided tax incentives for countertrade business and has licensed eight businesses for this purpose. In 1987 countertrade through Singapore totalled about US$500 million, and exceeded US$2 billion in 1990. Essentially, countertrade in Singapore is one where companies conduct arbitrage operations, based on their extensive marketing networks, and no doubt building on Singapore's premier position as an entrepôt port. Thus, where some trading companies are unable to provide suitable barter commodities, Singaporean countertrade companies act as brokers to find third parties to effect a transaction. In such arbitrage activities, therefore, countertrade serves to increase Singapore's commercially pre-eminent position.

Indonesian counterpurchase

Counterpurchase policies in Indonesia provide a rich source of material for study since it is stated government policy for all procurements. The intention is to alleviate foreign exchange difficulties and to reduce balance of payments problems. While Indonesian counterpurchase is a government responsibility trading through various state trading corporations, the onus is on the foreign exporter to seek out export commodities that comply with policy. Also in doing so, the foreign exporter must pay attention to additionality, that is, that the volume exported, other than oil, is over and above what would normally have been exported in that trade year. The duration of the contract obligation begins on signing the contract and ends on final delivery. Penalties are imposed equal to 50 per cent of that portion of the contract not fulfilled. However, there are also exceptions to counterpurchase requirements and these include sales with a value of less than US$15,000, sales financed by concessional credits, that is, credit with interest rates of less than 3 per cent per

annum, and sales financed by bilateral grants, joint ventures and professional services including military procurements.

Countertrade has covered a wide range of products and the contracts
have been sizeable. Indonesia countertraded with Romania for railway
cars (US$42 million), West Germany for commercial aircraft (US$90
million), boats (US$88 million) and locomotives (US$14 million). A
large multinational arrangement with 10 firms from six different countries was concluded in 1982 amounting to US$127 million for fertilizers
in exchange for a like amount of Indonesian goods.

Prospects for countertrade

Countertrade is a contradiction. The question may be asked as to whether
trade is to be used as a political tool or as a system for the efficient
allocation of resources. Countertrade is in a state of evolution and many
governments are constantly reassessing their own programmes. However, with the liquidity crisis in the world monetary system, the countertrade phenomenon may well continue into the twenty-first century given
concerns over unemployment, poverty and over-population.

There are only a small number of countries with an effective countertrade programme. This form of trading seems to depend on the quantity
of natural resources a country has at its disposal, its creditworthiness, the
nature of its political system and the willingness of the government to
use countertrade as a way of developing its economy. Examination of
the socio-economic performance of Indonesia has shown that it has been
able to reduce its outstanding foreign debt although it has yet to tackle
its current account deficit problems. Indonesia has by far the most
comprehensive and elaborate counterpurchase policy in place in South
East Asia. Countertrade in many of these countries suffers from the lack
of bureaucratic coordination which is necessary to take maximum
advantage of linked purchases. Also entrepreneurial skills are required
in the bureaucracy to heighten the profit motive. Goods offered for
countertrade will need to be made more saleable in world markets
through improvements in quality and economic attractiveness. These
steps may ensure that the restructuring of the economy in many LDCs in
South East Asia can proceed as planned. The main motivation for
countertrade appears to be the desire to increase competitiveness as
well as the *entrée* it provides to penetrate difficult markets. Moreover,
countertrade has the capacity to overcome currency exchange and credit
problems. Yet, there are risks associated with countertrade such as the

difficulties in both the valuation and the disposal of goods counter-traded. There is also the complexity and time-consuming nature of countertrade negotiations.

It is recognized that countertrade may not be the most desirable form of international trade. Countertrade sits uneasily with the concept of an open, cash-based, non-discriminatory trading system which the GATT and OECD aim to promote. While there are few laws in the US specifically concerning countertrade, the considered view is that it is trade distorting, protectionist and economically inefficient. The UK and Germany view countertrade as being potentially trade distortive but have refrained from intervention in private sector decisions. On the other hand, France has actively intervened to discourage countertrade. Certainly countertrade is not for every business. There is a set of characteristics which firms and their management require before they can be successful in countertrade. Those firms who engage in counter-trade tend to have had some experience of countertrade, have been established for a relatively shorter time and have senior managers who have spent time living in countries active in countertrade. Moreover, because countertrade is a high-risk activity, firms tend to be relatively large enterprises in terms of employees, manufacture standardized products for which there may be many competitors and adopt a plan-ning period approach to their business.

In the final analysis the growth in international trade by way of barter and countertrade in the foreseeable future will depend, on the one hand, on the price of major exports from LDCs and, on the other, on the world economy. When the price of LDCs' major exports stabilize and there is growth in world trade, then the consequent rise in foreign exchange reserves will enable many LDCs to return to their usual trading pat-terns. However, if the world economy were to develop a high degree of protectionism, and recession and other adverse economic ills were to continue, then it is suggested that countertrade will expand rapidly among LDCs. Whatever the scenario in the near future, countertrade will remain a feature of international business transactions (see Figure 4.2).

The transfer of technology

NIEOs Programme for Action

In the New International Economic Order (NIEO) of the 1970s, the Programme for Action demanded immediate redress on four fronts.

GARLIC DIPLOMACY

COMMUNIST North Korea and the running-dog capitalist South have agreed to break trade embargoes by allowing a deal involving 5,000 tonnes of garlic, Seoul officials have announced.

Garlic and spring onions are vital ingredients in *kimchi*, the ubiquitous fiery pickle side-dish which is considered an essential part of Korean diet.

Reuter quoted a South Korean Government official as saying a shipment of 1,000 tonnes of garlic was sent last month to the North from Inchon, west of Seoul, after permission was granted by the Unification Ministry.

He said South Korea had agreed to exchange 5,000 tonnes of garlic and 500 tonnes of onions for 1,300 tonnes of northern buckwheat and 100 tonnes of red beans — which sounds more like a food parcel exercise than a genuine trade deal.

Figure 4.2 Countertrade and international relations. The newspaper clipping gives a humorous but accurate view of countertrade as it is conducted even between rival neighbouring states
Source: *Sydney Morning Herald*, 11 November 1993

One such front included re-examining issues of industrialization and technology.

- Objective 15 of the NIEO involves the negotiation for a redeployment of the industrial productive capacity to developing countries. This issue proposes a shifting of the industrial capacity of developed countries to the Third World, especially those industries which have a high labour content, those industries which require natural resources and industrial processes requiring locally available raw materials.
- Objective 16 attempts to establish a mechanism for the transfer of technology. It is essential that developing countries have access to modern technology if they are to achieve their development objectives. The NIEO thus calls for greater access to technology through a review of international patents, the facilitation of access to patented and non-patented technology, the expansion of assistance to Third World countries for R&D programmes and the control of the import of technology.
- Objective 17 relates to the regulation and supervision of the activities of MNCs and the elimination of restrictive business practices. The proposal is for an international code of conduct so that such enterprises are prevented from interfering in the internal affairs of countries in which they operate, restrictive business practices are eliminated and that MNC activities conform to national development plans and objectives. Also the proposal encourages MNCs to assist in the transfer of technology and management skills to developing countries, to regulate the repatriation of profits accruing from off-shore operations and to promote the restructuring of their profits in developing countries.

The use of technology and sharing technology is important to developing countries. It may be argued that a principal weakness among most developing countries is both the lack of access to technology and their command of it. The acquisition of technology can determine growth and the capacity for growth. However, merely acquiring technology itself is insufficient since a country will only be able to benefit from additional technology if it can provide the 'welcoming structure' and a capacity to absorb these new ideas and techniques – the so-called 'absorptive capacity'. Such efforts towards a greater technological self-reliance need to be fully supported through various forms of international cooperation. Technology transfer may involve the sale of patents and equipment, the supply of know-how and processes and the dissemination

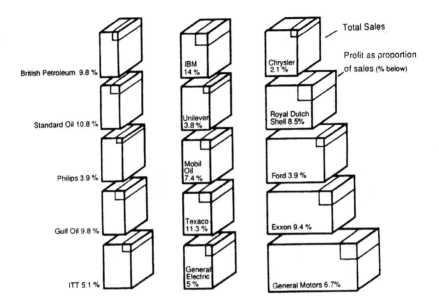

Figure 4.3 The multinationals: a select sample. The emergence of vast multinational corporations (MNCs) is of more than economic and industrial significance. The activities of MNCs affect and influence government policy and international relations. MNCs operate throughout the world and are examples of international cooperation *par excellence*. Many have annual sales exceeding the GNP of small developing countries.
Source: Adapted from *The Australian World Atlas*, London: George Philip, 1981: 21

of scientific and technological information through various means. The benefits stemming from such activities include the reduction of production costs, increases in production efficiency that are accompanied by the elimination of wastes, the reallocation of capital resources and improvements in the use of technology that result in higher productivity and improved techniques (see Figure 4.3). However, there are 'backwash effects' resulting from the transfer of technology. These include:

• the spread of international capitalist domination of domestic economies through the investment activities of private MNCs;
• the export of unsuitable and inappropriate science and technology protected by an exploitative patent and licensing system;
• the tactic of 'imposing' their products on fragile Third World markets

behind the screen of import substituting tariff barriers that inadvertently serve to protect the monopolistic practices of MNCs;

- the transfer of outmoded and irrelevant systems of education and university training to inappropriate international professional standards to countries where education is a key component in the development process;

- the 'dumping' of cheap products in controlled markets by industrial countries, thus upsetting the industrial process in developing countries; and,

- harmful international trade policies that lock Third World countries into primary product exports that result in declining international revenues.

Strategies are therefore required in order to minimize the debilitating effects of technology transfer and to focus more intensively on its positive aspects.

Transfer of technology strategies

The strategies required to implement technology transfer are dictated by particular needs. It appears that the needs are two-fold. There is a need first for technical change rather than the accumulation of capital before economic growth and sustained economic development can take place. The technology brought in by foreign investment is regarded as a potentially more valuable element than the capital inflow. But such industrial technology needs to be adapted to the particular condition and to particular endowments of various factors of production. Second, there is a need for trained engineers and scientists to make use of imported technology. Here the concern is also with the appropriateness of the technology, the terms on which it is bought and sold, its licensing and other aspects of foreign direct investments by MNCs in turnkey projects. Technology is transmitted in a number of ways, for example, through published literature, personnel exchange, imitation and copying. However, most of it comes with the sale of machinery and know-how, through training and technical assistance or through participation in the construction, operation and management of foreign firms in the developing country. Admittedly, technology transfer may also result from the reverse engineering of a product, the adaptation of products to suit local needs or simply direct copying.

Willy Brandt (1980) suggested that there might be eight ways in which more technology could be transferred to developing countries:

1 It is essential that information about technology should flow more freely between and within countries.
2 There should be greater support for technical assistance from UN agencies which provide an important channel for transmitting technology. Such agencies should be enlarged and made more effective particularly in giving more support to local research and evaluation of needs.
3 Special international support should be given to research into more efficient production, development and marketing to defend and improve the market competition of those raw materials mainly produced by developing countries, for example, rubber, jute, cotton and hard fibres, especially since most of these are threatened by synthetic substitutes.
4 Serious study should be undertaken of the implications and ways of coping with major technological breakthroughs in the industrial countries of the North, notably microelectronics, which may not only reduce the demand for labour in, say, North America and Europe, but also deprive developing countries in the South, like those in Latin America and Asia, of their comparative advantage in low-wage costs.
5 The situation should be changed whereby at present barely 1 per cent of spending on R&D in the North is specifically concerned with the problems of the South, whereas 51 per cent is devoted to defence, atomic and space research.
6 Aid agencies should make more use of local consultants and skilled people in preparing their projects and programmes.
7 Countries providing aid should give freedom to the recipients to make their own choice of imported technology. When aid is tied to donor sources it greatly limits choices and discourages local initiatives.
8 There should be more effective coordination in the many areas of technology which affect countries all over the world. Many of the technological challenges of today are but part of the basic problem of the survival and conservation of the world's resources.

One characteristic of the market for technology is that it is imperfect. This means that the differential between the price at which a seller is willing to sell and the price which the buyer has to pay will be very high because the technology has already been developed and its use can be very profitable. This is because it will be very expensive for the buyer to develop the technology from scratch. But at the same time the buyer can ill-afford to do without the technology. This situation places buyers in developing countries in weak negotiating positions because of their overwhelming dependence on imported technology from industrial countries. The cost of imported technology may be high in relative terms because of tie-in clauses and restrictions on exports. The high cost of imported technology may be assessed by the fact that it may cost relatively more to acquire the technology. Moreover, the profits flowing from the use of transferred technology may not be shared equitably between the supplier of the technology and the importing country. Put differently, the price of known commercial technology should be close to zero, especially if it is to be used in developing countries and is part of an aid–foreign investment package. For these reasons, the market for imported technology is said to be imperfect.

Technology imports are also bedevilled by a further double-edged problem. The successful transfer of technology will depend, on the one hand, on the willingness of the MNC to provide training and, on the other, on the willingness of the authorities in the host country to accept new technology. There is a mutual distrust on both sides and there are divided loyalties which make the problem doubly difficult. Given the impasse, it would seem that developing countries might do well to develop as rapidly as they can a local scientific and technical infrastructure. This may ensure some degree of technological autonomy. Moreover, as the country develops, the local subsidiary of the MNC may become independent of the parent company thus making it more imperative to develop local R&D capabilities. The added advantage, of course, is the use of local materials unavailable elsewhere and the production of goods ideally suited to the local market.

Pragmatically, some MNCs have entrusted local firms to devise products suited to local markets and this strategy appears to have been successful. Examples include the development of the Asia car concept, Spanish cosmetics and Mexican food. However, the creation of an innovative capacity within developing countries may seem to run counter to the interests of MNCs. The reality, however, is that international subcontracting, the manufacture of components and the local assembly of manufactured goods have worked successfully and quite profitably for all parties involved. In the long term therefore there is merit for developing countries to continue to strengthen their framework for R&D and to minimize their dependence on the transfer of technology from developed countries. In doing so, the capacity of developing countries to select

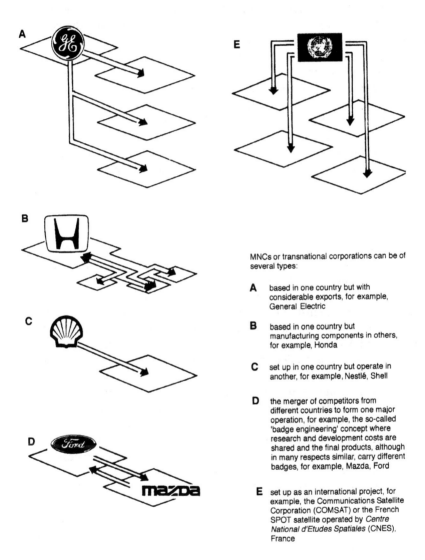

MNCs or transnational corporations can be of several types:

A based in one country but with considerable exports, for example, General Electric

B based in one country but manufacturing components in others, for example, Honda

C set up in one country but operate in another, for example, Nestlé, Shell

D the merger of competitors from different countries to form one major operation, for example, the so-called 'badge engineering' concept where research and development costs are shared and the final products, although in many respects similar, carry different badges, for example, Mazda, Ford

E set up as an international project, for example, the Communications Satellite Corporation (COMSAT) or the French SPOT satellite operated by *Centre National d'Etudes Spatiales* (CNES), France

Figure 4.4 Multinational corporations and the transfer of technology

technology that is most appropriate to their immediate and long-term needs will be enhanced (see the various types of MNCs and the possibilities of technology transfer shown in Figure 4.4).

Here, it may be apposite to distinguish between the various terms to describe technology. Alternative technology is the ecology, self-sufficiency and smallness-oriented version of technology that has been proposed to humanize western industrialization. Intermediate technology, a term favoured by Fritz Schumacher (1973) *Small is Beautiful*, aims to fill the gap between traditional and modern technology. Appropriate technology is an umbrella concept for all forms of small-scale hardware and software which are suited to the economic conditions of developing countries.

Appropriate technology

The technology transferred by an MNC may be inappropriate to the needs of developing countries. This may be because of the size and scale of the production technology, the production equipment and local conditions and needs. For instance, Moroccan mining output may be too small for it to install blast furnaces that could deal with a mixture of ores. The rate of assembly of motor vehicles in Morocco and the Côte d'Ivoire may not be high enough to make it possible to install production lines for a single model and vehicle type. Thus, in order to operate cement works in Kenya a choice has to be made among the capital equipment available to determine what would be suited to the raw materials available in Kenya. In production technology, MNC tyre manufacturers in Morocco and Greece have changed their methods of manufacturing rubber to make their tyres more durable in local conditions. The highest degree of adaptation of technology and processes is in consumer goods such as pharmaceuticals. This is because commercial reality in terms of consumer habits as well as differences in climate play a vital part in their adoption and use.

Technology may be inappropriate because it costs too much per work place, creates too few jobs, involves an excessive scale of production, leads to over-urbanization of capital cities and produces over-sophisticated products that do not meet the basic needs of the poor. In contrast, the technology that produces wide-ranging employment opportunities at least cost, produces goods that meet basic needs, makes use of local natural resources and is consistent with ecological principles that take care of the environment, is considered appropriate. For technology to be appropriate it needs to provide efficient levels of productivity when compared to alternatives.

Appropriate technology does not necessarily prescribe any particular type of technology, much less whether these should be the latest or most sophisticated. Appropriate technology also means that the choice of technology has been a deliberate one taken in the knowledge that it can and will affect the character and direction of development. Such appropriate technology can include cheaper sources of energy, simpler farm equipment, techniques in building and manufacturing that save capital, small plant size and scales of operation that permit the dispersal of activities and those that take account of the special nature of problems in each case.

Appropriate technology is thus technology that makes optimum use of available resources and also enhances these resources. However, the appropriateness of the technology itself begs the question – appropriate for whom, appropriate for where and appropriate for when? These questions suggest that there may be a number of alternative and competing criteria either implicitly or explicitly to gauge 'appropriateness'. These include:

- maximization of output;
- maximization of availability of consumption goods;
- maximization of the rate of economic growth;
- reduction in unemployment;
- regional development;
- reduction in balance of payments deficits;
- greater equity in the distribution of income;
- promotion of political development including national self-reliance; and,
- improvement in the quality of life.

In practice there is a general tendency for MNCs to use capital-intensive techniques of production even though there is abundant labour and where labour is the cheaper factor. This may be because the imported technology is perceived to be appropriate to the relative needs and availability of manpower skills and other resources. While a more labour-intensive technology may be used, it may be less economic given the loss of economies of scale or because skilled labour or supervisory–management staff is unavailable. Nevertheless, labour may be used more intensively in ancillary processes such as transport and handling and the construction of buildings. It should also be noted that inappropriate technology is not confined to industrial processes. For example, in African agriculture the research efforts to optimize physical yields per hectare reflect the background of European advisers or training. Optimizing yields in itself may not be the appropriate criterion given that traditional African agriculture may be constrained by seasonal shortages of labour rather than the scarcity of land and social practices.

Equally, the alternative argument also applies in that large-scale capital-intensive plants can be appropriate in the right context. Oil refineries or bulk fertilizer plants provide examples since these capital-intensive investments may be the only possible way to achieve the desired quality or because they represent the most effective use of capital. In these cases the technology is appropriate since it is the most economical use of a country's natural resources and its relative abundance or lack of capital, labour and skills. In addition, such technology furthers national and social aspirations. These are practical requirements since the technology will be sound, economical to use in the light of available alternatives and socially acceptable in the context of the local culture and traditions.

It is believed that the appropriateness of technology flows from the natural consequences of the product cycle. R. Vernon's (1973) *Multinational Enterprises* suggests that there is an explanation for the multinationalization of production. He postulated that there are three phases in the product cycle. First, with a new product, the technology is not yet stabilized since it incorporates recent scientific discovery. The price is high and market limited and the product is sold to consumers who are able to pay for it. The second phase of the cycle leads to the standardized product. Here the technology is better known and the market much wider so that it is now within reach of consumers in industrial economies. The third phase is the mature product which is manufactured using simple technology and the costs are greatly reduced. It is this stage which contributes to the industrialization of the Third World with the transfer of sectors producing goods that have reached maturity. Here may be seen the basis of the arguments for the new international division of labour and the more general phenomenon of globalization of production.

It is clear that developing countries will have great difficulty in participating in the initial two stages of the product cycle. However, there can be more general participation by all in the third stage because the technology by now has become more appropriate and more amenable to widespread use and adoption. Moreover, the acquisition of technology at arm's length will remove the threat that developing countries will suffer from the adverse effects of the activities of MNCs. South Korea and Taiwan are examples of countries who have managed to get the best out of technology without suffering its worst effects.

In general, in so far as MNCs are concerned, the record has been ambiguous. MNCs have been known to make big profits by exploiting the cheap 'disciplined' labour and raw materials, and have transferred inappropriate technology which aggravates mass unemployment or 'dumps' products that are harmful to health within an unregulated consumer system. The infamous examples of infant milk formulas, pharmaceuticals and drugs and pesticides speak eloquently of such tactics. Developing countries offer generous investment incentives to MNCs to the extent that there are sub-minimal safety standards and low- to non-existent environmental standards. The Bhopal tragedy in India offers a graphic reminder of lax standards. In turn, the élites in developing countries have adopted the consumerist model of modernization and one which ensures that social inequality becomes more pronounced, the environment degraded and urbanization accelerated. The modernization model in which MNCs transfer useful capital and technology to developing countries and offer markets and employment has to be matched against the model in which MNCs take wealth out of developing countries, provide limited jobs with over-capitalized intensive technology and exploit cheap labour and other resources. In many developing countries MNC activities are confined to free trade zones, institutionalized enclaves that have few links with the rest of the economy. The profits are invariably not taxed, having been given tax holidays, and the labour force is lowly paid, with little or no opportunities for unionization. There is little or no back- ward linkage between the MNC and the local economy, so that the immediate impact of the MNC on the development of the local economy is small. Overall, therefore, the impression is that the MNC is a poor vehicle for development unless its activities are closely regulated to accord with the principles and goals of develop- ment set out by the developing country.

Case study E

Mexico's *maquiladores*

In the run-up to the twenty-first century many developing countries have come to realize that growth and development can be achieved by participating in the production of a common world product. This globalization of the manufacturing process is unprecedented and will appear to be the trend in the near future.

Mexico is liberalizing its economy after a long history of isolation, especially from its neighbours, and decades of pursuing import substitution strategies. As part of its efforts to liberalize trade, tariffs have been reduced drastically and the use of official prices for valuation purposes has been discontinued. In addition, many new sectors have now been opened up to foreign participation and a process of privatization of state-owned enterprises has been put in place so that nearly two-thirds will be privately owned. The US market is becoming more closely integrated with Mexico as firms take advantage of the Mexican reforms. One spatial manifestation of such economic integration may be observed in the border regions.

The Mexican international economy is made up of two parts, its border region and 'other' international production. The border region, delimited in 1966, consists of a 20 km strip the length of the frontier with the US where free trade prevails. Companies were granted the right to create 'in-bond' plants known as *maquiladores* which have the right to import capital equipment and raw materials with the proviso that all manufactured goods were exported. In a *maquiladore*, a US or other foreign firm sends components, raw materials and equipment duty-free to Mexico where the final product is assembled and then exported. Moreover, the customs duties on products re-exported to the US are only levied on the value-added in Mexico. This structure has thus allowed US and other foreign companies to take advantage of lower labour rates in Mexico.

Maquiladores provide a means for implementing import substitution policies in Mexico. There were 73 plants employing 16,000 workers in 1967 which had grown to 15,000 plants and employing close to 250,000 workers by the 1990s. The value of

Case study E (*continued*)

the exports from the *maquiladores* is estimated to exceed Mexico's manufacturing exports from the rest of the economy.

Intermediate economies and relationship enterprises

The stage has now been reached where perversely some NICs may prefer to remain as such fearing that to do otherwise may mean the loss of access to international development funds and aid generally. Countries like Malaysia and Thailand come within this category. On the other hand, the more successful Asian nations such as South Korea, Taiwan, Hong Kong and Singapore are in an anomalous position of being developed in some respects and developing in other respects. Such economies have been described by economists as 'intermediate economies' in that they have characteristics that are in common with both the more developed industrial countries and the developing nations. MNCs operating within such economies adjust their production technology as well as products to coexist with this host economy.

An indicator of economic development is that as the country's per capita income and stock of human capital rises, there is a shift from being a net importer of technology and skills to being a net exporter. In this transformation it may be that during a transitional stage the economy may be able to export a narrow range of highly specialized technological products while at the same time remaining a net importer of technology in aggregate terms. The suggestion is that in the 1990s some successful developing countries may have reached this phase in their development. Any one of the Four Dragons would make an ideal case study for examining issues associated with early stages of technological exports. Such exports comprise the international flow of basic technical know-how and management services, and other skills. The technology includes the technological components of foreign investment and overseas construction, licensing and technical agreements, consulting services and other commercial agreements.

One pressing issue that impinges on the intermediate economies concerns their spectacular growth and continued impressive performance when compared with other developed economies. The Four Dragons of Asia, for example, have been able to sustain high growth rates compared to the richer industrial countries over the last 20 to 25 years. Their per capita incomes have nearly quadrupled, poverty rates have shrunk and if such growth is sustained the East Asian region may overtake North America as the world's dominant economic region. How and why this has come about is the subject of contentious debate and divided views.

Neo-classical economists claim that the most successful Asian economies were the ones that got their macro-economic basics right – combining low inflation, sound management of internal and external debts and keeping government intervention at a minimum. On the other hand, detractors point to the industrial policies, of say Japan, which show highly restrictive policies on capital outflow and investment tax regimes and the repressed interest rates in South Korea, Taiwan and Singapore. A World Bank Report (1993) entitled *The East Asian Miracle* suggests that each of the successful economies has ensured that their economic basics were under control, including low rates of inflation, competitive exchange rates, strong financial systems and careful debt management. There was also high investment in human capital in Asian economies in sharp contrast to Latin America and Africa. The successful Asian economies practised some form of government intervention of the economy to stimulate growth although it may be difficult to state categorically that government intervention directly contributed to rapid economic growth of the countries concerned.

Analogous to the concept of the product cycle discussed in the previous section, the growth of the Asian economies has been described as a 'flying geese' pattern. This pattern refers to the tendency for Asian economies to move up the manufacturing services chain in sequence from labour-intensive to capital-intensive manufacturing and then into technology-intensive manufacturing and finally into services-related activities. In the case of South Korea and Taiwan, the emphasis has been on R&D with the government taking a greater role in Taiwan whereas in South Korea greater reliance has been put on the larger conglomerates known as *chaebols*. In the high-wage economies of Hong Kong and Singapore there has been the development of niche markets such as computer component manufactures and financial services, especially offshore banking and insurance. There is increasing specialization in manufacturing with each of the Asian nations concentrating on their comparative advantages so that Hong Kong and Singapore are seen as having advantages in banking, insurance, electronics and pharmaceuticals, whereas South Korea and Taiwan may have advantages in steel, paper and textiles. Similarly, Thailand, Malaysia,

Indonesia and the Philippines may have comparative advantages in textiles, fertilizer and cement but have disadvantages in banking, transport and precision equipment.

Apart from the impressive growth patterns of the Asian economies the context of the transfer of technology and the globalization of manufacturing has produced new forms of business relationships. Around the world there has grown up so-called 'relationship enterprises', alliances of otherwise independent MNCs. Examples of such enterprises include Boeing–Airbus, McDonnell Douglas–Mitsubishi–Kawasaki, Intel–Microsoft, Mazda–Ford and General Motors–Toyota. The advantages stem from the potential to undertake large projects which are well beyond the resources and capabilities of a single corporation. Also, such transnational conglomerates may be less subject to national commercial trade practice laws and together may have greater access to both national and global capital markets.

The problems, however, of such relationship enterprises include those of coordinating management of the diverse arms of the corporation, the difficulties of setting up commercial strategies let alone enforcing compliance of decisions and, more importantly, the difficulty of both apportioning and distributing returns to the various parts of the conglomerate. Of course, all these developments have come about because of economic imperatives confronting the global economy such as the scarcity of capital as well as the emergence of new players including the Asian economies described previously. There is thus an opening up in locational opportunities for MNCs globally. The change is from adversary and competitor to cooperation and facilitation – a model reflected in the promotion of freer trading and investment regimes.

In the Asian context relationship enterprises take a different form. A phenomenon much discussed but seldom put within the context of trade, development and MNCs is the role of overseas Chinese business networks. The importance of 20 to 30 very large business groups, mostly owned by ethnic Chinese, is a feature which dominates the commerce and industry in Asia. For example, the Lippo Group in Indonesia and Hong Kong, the Salim Group in Indonesia, the Sophonpanich Group in Thailand and the Kuok Group in Malaysia are prominent players within their respective countries as well as in other parts of Asia. Overseas Chinese networks are so named because they are dominated by expatriate native-born Chinese or their offspring living outside China. The aggregate wealth of the 40 million or so overseas Chinese has been estimated conservatively to be about about US$250 billion.

The networks are held together by capital flows and joint ventures as well as by marriage, and by commonalities in cultural and business ethics. Such networks have been formed in trade, finance, real estate, processing and assembling and have grown in sophistication from the informal one-person patriarch decision-maker to the more formal board of directors using western trained executives and modern business management techniques. Such networks produce linkages with international capital although there still remain vestiges of family relationships and obligations among relatives.

The growth of Asian MNCs is seen as the result of a need to service Chinese clients in Asia. In Hong Kong and Singapore, for instance, Asian MNCs are present in banking – Hong Kong and Shanghai Bank, Standard and Chartered Bank, which have extensive operations in Asia – and telecommunications – Singapore Telecom, which has expanded aggressively into the Asian region. Already, strategic alliances are being formed such as the Mitsui Group's joint venture with overseas Chinese in Thailand. The Lippo Group in Indonesia has moved out of its traditional financial business into cognate industries such as investment banking, securities, property and information technology. In insurance, the Australian Colonial Mutual Group has entered into a tripartite joint venture with Jardine CMG Life Group (HK) and PT Astra – an association which reflects the attitude and favourable perception European firms now have of Asian MNCs. Rather than a debilitating competitive model, the trend is one of joint ventures and strategic alliances between European and Asian MNCs.

Case study F

Transfer of know-how and MNCs in Asia

Towards the end of 1993 the *Far Eastern Economic Review* conducted a survey of 200 of Asia's leading companies to identify the factors that make a company an Asian leader – factors that went beyond mere sales and profits data. The review 200 comprised 90 MNCs from outside Asia that conducted business in the region and the 110 companies based in Asia. To be a leader a company cannot be a niche player, known and well-regarded only by the few it serves. Instead it must be recognized for its achieve-

---

Case study F (*continued*)

Table F.1 Leadership quality indicators

Leadership qualities

Respondents ranked companies by these attributes:
  High quality of service or products
  Management has long-term vision
  Innovative in responding to customer needs
  Financial soundness
  Companies that others try to emulate

---

ments and leadership qualities. More than 4,000 leading professionals were surveyed for their views on five attributes that provide a measurement of corporate leadership. The leadership qualities are shown in Table F.1 while the results from the survey are given in Table F.2.

Until very recently, Asia, apart from Japan, was a low-cost production site. There were no markets of any substance given that Japan was difficult to get into. However, by the mid-1990s attention was increasingly focused on Asia, whether it be the market for software or sewers – market demand is growing more quickly in Asia than anywhere else. The leading MNCs in Asia sell everything from soft drinks to cellular telephones, emphasizing the importance of consumer products. Perhaps this is a reflection of Asian people wanting to be part of the world and aspiring to be modernized. But these are not recent phenomena as the process has gone on for a long time. Coca-Cola first set up its bottling plants in Asia in the 1920s. IBM opened in China in the 1930s and Motorola made black-and-white televisions and vacuum-tube radios in Asia after the Second World War. Increasingly, however, there has been a change in emphasis, from merely exploiting low-cost production sites to one where there is a long-term commitment and a genuine transfer of knowledge, skills and technology.

The transfer of technology is seen everywhere but manifests itself most clearly in terms of 'localization' of executive personnel, of production and local demands and competing locally. The challenge that every MNC faces in Asia is how to transfer more executive skills to Asian counterparts. Staff rotations in Asia and overseas is one means – a management style that the US-based electronics firm Hughes Corporation has used, as has General

Table F.2 The *Far Eastern Economic Review* results of a survey of 200 multinationals, 1993

Rank	High-quality services/ products	Companies that others try to emulate	Innovative in responding to customer needs	Financial soundness	Long-term vision	Overall rank
1	Rolex	Coca-Cola	McDonald's	Coca-Cola	Coca-Cola	Coca-Cola
2	BMW	McDonald's	American Express	Citibank	McDonald's	McDonald's
3	Kodak	Walt Disney	Apple Computers	Deutsche Bank	Microsoft	Kodak
4	Rolls-Royce	Microsoft	Walt Disney	Bank America	Walt Disney	Walt Disney
5	Xerox	IBM	Citibank	Esso/Exxon	AT&T	Rolex
6	Gillette	Xerox	DHL	Chase Manhattan Bank	Boeing	Xerox
7	Adidas	Apple Computers	Microsoft	British Petroleum	Hewlett Packard	Nestlé
8	Nestlé	Daimler Benz	Coca-Cola	McDonald's	Motorola	Motorola
9	Motorola	DHL	Nike	Royal Dutch/Shell	Du Pont	Boeing
10	Boeing	Rolex	Motorola	American Express	General Electric	AT&T

Case study F (*continued*)

Motors and McDonnell Douglas. The localization of production has helped companies like Boeing and Coca-Cola to expand and dominate in Asia. For example, China is Boeing's largest overseas customer with almost one in every seven aircraft it sold in 1993 going to China. While US and European MNCs have had an advantage over their Asian rivals, in the 1990s Japanese, South Korean and Taiwanese firms have increased their investments throughout Asia. Other MNCs which have not had a presence have begun to invest in Asia. AT&T is focusing on China by planning to establish world-class laboratories and microelectronic facilities. In this task AT&T is facing tough competition from entrenched competitors such as Alcatel which has been building telephone switching equipment in China for more than a decade. However, some MNCs are less keen to participate in this dynamic market unless changes are brought about to protect intellectual property rights. For example, Microsoft and Walt Disney Enterprises have been hesitant about investing heavily in China because of their fears of computer piracy, fakes and infringement of copyright.

To illustrate the successful transfer of technology and trends in skills transfer two examples are discussed below. In 1993, Coca-Cola was bottled at 13 sites across China including five state-owned plants. Another 10 plants will be built by 1997. The pattern used by Coca-Cola is to team up with local partners for everything from bottling to glass-making and to use local ingredients and management as far as possible. The bottling operations are tied closely with partners who have good distribution networks and strong financial resources. It is crucial that local bottlers use local networks because otherwise transport costs would make the drink uncompetitive in the market. The two ingredients of a strong local distribution network and heavy localization ensures that the instantly recognizable Coke ribbon can increase its market share even further. Subsidiary industries such as plastic suppliers and drink-can manufacturers also stand to benefit when associated with Coke manufacturers. In 1993, the Coca-Cola system in Asia employed 85,000 people but only 1,600 (1.9 per cent) were paid by Coke, the rest being employed by bottlers.

A second example of MNCs supporting local industry and developing local expertise is that of the Swiss-based food

Case study F (*continued*)

company Nestlé. The Nestlé approach is to integrate itself into the local economy – a tactic which it has used successfully throughout the world. For example, Nestlé sent agronomists to encourage coffee cultivation among the Dai tribe in the mountainous southern Chinese province of Yunnan. In the northern province of Heilongjiang, Nestlé has promoted the local dairy industry by cornering the local milk market. The development of local services is repeated in Thailand, India and Indonesia where the indigenous dairy industry is being encouraged and in Malaysia where it is encouraging coffee cultivation. This methodical grass-roots approach in developing local raw materials is complemented by heavy investments in manufacturing and marketing. Nestlé operates in virtually every country in Asia except North Korea and Vietnam. Nestlé's revenues in 1992 from Asia totalled US$4 billion, with Japan alone accounting for about one-third of Asian sales (US$1.77 billion). Nestlé also has a strong position in Malaysia where it posted sales of at least US$510 million.

Source: See *Far Eastern Economic Review*, 30 December 1993 and 6 January 1994.

Conclusion

The term globalization, to describe the conduct of business across national boundaries and the taking in of a regional and world-wide perspective, is both a buzz-word and a trend in international trade. The underlying motivation among American, European, Japanese and Asian MNCs is driven by the need for continuing market access as well as for circumventing all sorts of barriers to trade. This chapter has examined the rationale and persistence of countertrade as an alternative means of international trade. Whether this will persist in the near future will depend on individual countries and their need to go outside mainstream trading channels. Apart from trade, many developing countries seek to increase their productivity through technology. This transfer of technology is effected through imported capital goods and other inputs.

Such technology may be bought through licensing agreements, transmitted through direct foreign investments, labour movements or contacts with foreign buyers. In an open global market, the supply of new products and processes will be further enhanced. However, the rapid rate of technical change in microelectronics, telecommunications and biotechnology is creating a very competitive and complex world. Developing countries therefore need to be able to adopt and to adapt to these new technological changes. This need is the result of being integrated with the global trading system. That many developing countries are unable to fulfil such demands suggests that they lack the capacity to absorb technology because either these countries do not have enough skilled people to organize the use of foreign investments or lack the physical infrastructure to assist in technological progress. The next chapter examines the globalization of trade and the integration of regions.

Key ideas

1 Countertrade, an alternative form of international trade, involves various forms of exchange of goods and services. This new mercantilism is responsible for bringing about the favourable balance of trade of many less developed countries (LDCs).
2 Many LDCs engage in countertrade because of a rise in protectionism of markets, a decline in confidence in the world trading system, growing indebtedness and the need to penetrate new markets.
3 The use of technology and sharing technology is an important means of economic development. Unless developing countries have access to technology and have a command of it, economic growth and progress will be slow in coming.
4 Technical change and a well-trained workforce in science and engineering will make the transfer of technology more effective.
5 Technology is deemed appropriate when it meets the needs of developing countries in terms of the size and scale of the production technology, production equipment and local conditions of use. The criteria for appropriate technology include the creation of employment, meeting basic needs, use of local resources and ensuring that it is consistent with ecological principles. The product cycle is one manifestation of appropriate technology transfer.
6 Some NICs have 'graduated' to developed status and among these are intermediate economies which have begun to export technology and

skills. The growth of some Asian economies has been described as analogous to the 'flying geese pattern' where such economies progressively move up the manufacturing services chain and from labour-intensive to technology-intensive manufactures. Indigenous Asian MNCs and relationship enterprises assist this progress.

5
Global trade reform

Introduction

Development may be thought of as an improvement in the quality of life (World Bank 1991). In the developing world, a better quality of life is based on more than higher incomes as it also encompasses better education, higher standards of health and nutrition, less poverty, a cleaner environment, more equality of opportunity, greater individual freedom and a richer cultural life. To achieve higher incomes necessarily requires higher productivity which itself is based on technology, investments in human and physical capital and international trade. The quality of the economic environment in which the development takes place is assessed by the extent to which markets are distorted, including foreign exchange restrictions, unrealistic price systems and restrictions on trade.

The 1989 Nobel Prize winning economist Trygve Haavelmo has argued that trade 'moulds the world's production structure in the direction of pressing for markets and advertising etc. of all kinds of less basic/necessary goods and gadgets. It tempts the poor countries to ruin their own valuable resources in return for imported machinery and consumer goods' (cited in Harris 1991: 3). Trade in developing countries should be based on a programme of what a country's economic activity should look like in the long run.

There are of course other external factors which influence development. In broad terms these factors include the terms of trade, international interest rates and capital flows and growth in industrial countries. Any plan to assist development would require a concerted effort from all involved. Industrial countries will need to lessen restrictions on trade, rethink macro-economic policy and ensure the success of the Uruguay Round of the General Agreement on Tariffs and Trade (GATT). Industrial countries and multilateral agencies will need to increase financial support to developing countries, encourage policy reform and promote sustainable growth. For their part, developing countries will have to invest in people, improve the investment climate, open their respective economies to international trade and investment and ensure that domestic macro-economic policy is correctly formulated.

These steps therefore provide the broad basis for sustained growth in productivity which may ensure long-term development. The issues revolve around the questions of whether policy interventions to promote free trade will be more beneficial than a free market. The free trade versus protectionist market debate looms large. This chapter discusses global trade reform – a topic of considerable interest to every trading nation around the globe. Yet, there are conflicting views as to the necessity of global trade reform, its implications and its impact on the social and natural environment. The first part of the chapter discusses the arguments for free trade as against protectionism. Various trade weapons and policy interventions are canvassed. In the second part, the need for global trade reform is analysed in the context of the growth of regionalism and trading blocs which have mushroomed in different parts of the world. Then the third and fourth parts discuss the GATT as the mechanism that controls world trade and how its various 'rounds' have brought about different results. The background to the General Agreement on Trade in Services (GATS) as an offshoot of the GATT deserves separate treatment because it has the potential to determine the outcomes of bilateral trade negotiations internationally. The fifth section addresses the impact of international trade on the environment, and concluding section briefly examines the new world order of Ekins (1992) and the way forward with respect to trade.

Free trade and protectionism

Trade offers developing countries important external relationships. The classical view suggests that international trade will be beneficial to every society. However, sceptics argue that the gains from trade are unlikely to be significant because developing countries have a heavy dependence on a limited range of very specialized primary products. These products are highly vulnerable to changes in international demand and hence subject to violent price fluctuations. A radical view is that international trade itself is the cause of the underdevelopment among many nations. This is because the prices received for primary products relative to the prices paid for imports have declined steadily. The terms of trade seem always to be working against developing countries. It is paradoxical therefore that the theory of comparative advantage suggests that trading partners will benefit from the exchange of goods and from specialization. Furthermore, productivity theory adds that gains from trade will accrue because of wider markets and increasing specialization and because trade has the effect of encouraging technical innovation and raising the levels of skill among workers. Trade also creates new industries and higher productivity.

Trade theories illustrate the fact that free trade is best because it has the effect of maximizing world economic output. The theory is that in a *laissez-faire* environment all participants will benefit from specialization and exchange. Here free trade may be defined as that trade in which goods can be imported and exported without any barriers in the form of tariffs, physical quotas or any kind of restriction. Trade which acts as a substitute for factor movements will benefit all participating nations. If trade was not free then nations would be diverting their resources and those factors of production in order to produce additional goods. These additional goods may be more expensive because that country may not have a comparative advantage in those products. It follows that if every nation were to pursue this policy, then ultimately the total production for all nations will be less than would otherwise be the case if the benefits of comparative advantage were pursued. Any deviation from free trade will reduce production and leave less to go around. This hypothesis is also grounded on the fact that the costs of transport do not wipe out the production benefits of specialization. However, another issue that needs to be addressed is that while there may be increased production, how will the increased wealth be distributed, and this is a political issue. Also, while free trade is efficient from a production point of view, the theory neglects to consider whether such trading is fair and equitable.

International trade and the conditions under which it takes place are determined to a large degree by dominant trading countries who are also the most developed economies. In terms of the history of international trade, the conditions have swung from one extreme to another. In the mercantilist environment of the eighteenth and nineteenth centuries the prevailing free trade notion was that what was good for Britain was good for the world. Free trade established an inequality between nations and competition was agreeable only after Britain had established its pre-eminence as the most advanced industrial nation. This artificial 'division of labour' was broken when other countries adopted protectionist policies to counter the effects of Britain's pre-eminence.

In general, the second half of the nineteenth century produced conditions in which free trade was the norm. This suited the economies of north-western Europe and was a period during which major areas of export specialization were developed. The rich nations were able to specialize in heavy industry and capital-intensive goods while the poor and developing countries concentrated on primary products. However, by the first half of the twentieth century protectionism was the norm in the developed world. This prevented specialized exporters from using their export income to diversify into the various types of processing and manufacturing industries since these were protected by the developed countries. There was an enormous reduction in the volume of trade following the adoption of protectionist policies during the period between the two world wars. This was deemed the 'cost' of protectionism. However, after the Second World War, developed countries began once again to discuss the need for trade liberalization, and as the beginning of the twenty-first century draws nearer the rhetoric is towards free trade within trade blocs and regional groupings.

While the decision to trade is made largely by individual firms and individuals, these decisions are influenced by a complex system of international conventions and arrangements. From a governmental point of view, the underlying agenda includes:

- the stimulation of international trade to benefit the nation. To this end, governments may also sponsor trade promotion and provide export subsidies;
- the protection of national interests by means of tariff barriers such as a tax on imported goods, import quotas by restricting the quantity of imports over a stated time period, exchange control through the rationing of foreign exchange and foreign currency movements and trading through state-owned enterprises; and,
- the promotion of trade within regional groupings through the formation and participation in free trade areas, the use of international currency blocs such as the dollar and sterling areas, and more

recently the yen, and trade promotion of world-wide trade reform through international bodies such as the GATT and the United Nations Conference on Trade and Development (UNCTAD).

The commercial policies of many developed and developing countries include those which hinge on 'trade protection'. These protective policies are designed to foster industrial growth, diversify exports, create employment and achieve other development-oriented strategies. Such policies include protective tariffs, physical quotas or quantitative restrictions, subsidies and commodity agreements. A tariff *ad valorem* (according to value) is a fixed percentage tax on the value of an imported good which is levied at the point of entry into the importing country. The term customs duties describes accurately these imposts. Quotas usually are a physical limitation on the quantity of any item that can be imported into a country, for example, the agreed number of vehicles imported per year. Such quotas are considered 'non-tariff' trade barriers (NTBs), but nevertheless they are barriers to free trade. Other examples of such NTBs are sanitary requirements for imported meats and dairy products and unnecessarily strict health and safety standards. Japan is notorious for imposing very strict 'standards' and being very fastidious in terms of what imported goods are allowed in to Japan. The length of the rice grain, its 'stickiness', the firmness of fresh fruits are some examples of attempts to reduce the volume of imports. In an indirect way, the volume of such trade is reduced by imposing more and more stringent standards (see Figure 5.1 for an illustration of the various forms of non-tariff barriers).

Quotas and other restrictions can also form a part of the so-called voluntary export restraints (VERs) in which trading countries agree to export a certain amount to the other country in exchange for some other consideration. Invariably most developing countries have no option but to agree to these conditions. However, this restraint also applies among industrial countries. In the 1980s, VERs imposed by the US on Japanese motor vehicles forced Japanese car makers to set up plants in the US in an effort to circumvent the trade barriers. VERs have been described by some as a pervasive and pernicious form of non-tariff barrier and are regarded as discriminatory. As noted, VERs are usually imposed by coercion, and are achieved by threatening anti-dumping or countervailing actions.

Other countries provide subsidies for their export industries so that the country's products are more competitively priced in the international

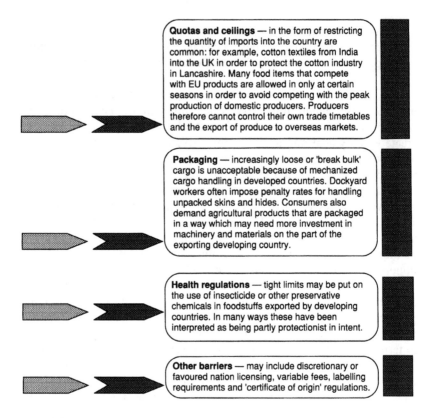

Quotas and ceilings — in the form of restricting the quantity of imports into the country are common: for example, cotton textiles from India into the UK in order to protect the cotton industry in Lancashire. Many food items that compete with EU products are allowed in only at certain seasons in order to avoid competing with the peak production of domestic producers. Producers therefore cannot control their own trade timetables and the export of produce to overseas markets.

Packaging — increasingly loose or 'break bulk' cargo is unacceptable because of mechanized cargo handling in developed countries. Dockyard workers often impose penalty rates for handling unpacked skins and hides. Consumers also demand agricultural products that are packaged in a way which may need more investment in machinery and materials on the part of the exporting developing country.

Health regulations — tight limits may be put on the use of insecticide or other preservative chemicals in foodstuffs exported by developing countries. In many ways these have been interpreted as being partly protectionist in intent.

Other barriers — may include discretionary or favoured nation licensing, variable fees, labelling requirements and 'certificate of origin' regulations.

Figure 5.1 Different forms of non-tariff barriers

marketplace. Farm subsidies are an example where farm produce seems cheaper and may have the effect of undercutting the competitors. The 'dumping' of agricultural surpluses or the stockpiling of commodities to maintain a price level depresses commodity earnings of developing countries. The control of foreign currency and foreign exchange is one way governments are able to restrict free trade between private individuals. This may have the effect of prompting individuals to circumvent the system by various means such as countertrade which was discussed in Chapter 4. Indirectly, by easing such controls, governments may attract multinational corporations (MNCs) simply by having no restrictions either in the repatriation of profits from host countries or in the types of currency repatriated. The benefits are therefore two-way – to the host nation and to the MNC.

The arguments for protection may now be summarized. Developed countries use various means of protection and these are designed to achieve the following:

- to protect against the use of cheap foreign labour and thus protect local industries;
- to improve the terms of trade. A tax will make the imported product more expensive in the local market. In order that it be sold in the local market, foreign exporters will be forced to lower the price of the product; and,
- to divert demand from foreign to domestic goods so as to shift a country's unemployment problem offshore on to foreign countries.

Developing countries use protection to achieve similar goals but with slightly different emphases:

- to protect 'infant industries'. Infant industries are usually set up behind protective tariff barriers as part of a policy of import substitution. The thinking is that once the industry is no longer an 'infant' the protective barriers are supposed to disappear. But in reality, these barriers often persist long after the industry has matured;
- to protect government and private investments. Through the use of subsidies, the effect is to make local goods more competitive both overseas and at home. At the same time protectionist policies are beneficial because they introduce economies of large-scale production offered by technology; and,
- by ensuring that domestic consumers purchase locally produced goods, the apparent effect is to substitute imports and in so doing save foreign exchange and assist in the balance of trade.

Clearly these policies are designed to protect national interests and are crucial for attaining a developed status. While these policies do not do much to stimulate international trade either quantitatively, in terms of the volume and variety of goods traded, or in promoting specialization and benefits from comparative advantage, the trend appears to be towards the promotion of trade within regional groupings. While in strict definitional terms the trade is international by its very nature, the advantages of a free and unhindered movement of goods are lost because seemingly there are built-in mechanisms that tend towards trade preference and favoured nation status. With discriminatory trade rules in place developing countries are squeezed further by the trend towards trading blocs.

130 Trade, aid and global interdependence

The net impact of tariffs and non-tariff barriers on developing countries will be a reduction in foreign exchange earnings. Tariffs, for example, increase with the degree of processing – the more highly processed, the higher the charge: compare the market price for groundnuts and groundnut oil. This is a disincentive to the developing country to diversify into secondary export industries, for example, turning out raw cotton instead of shirts and cloth. The high tariffs therefore are a restraint to industrial expansion and diversification. The stringent sanitary rules and regulations in force in developed countries have the effect of lowering the effective price developing countries receive for their exports. In other words, it lowers the terms of trade of developing countries. A further impact is that such rules effectively reduce the quantity exported and this is reflected in the lower foreign exchange earnings of developing countries.

Earlier it was suggested that tariff barriers are put in place in order to prevent manufacturing industries losing out to the low-cost labour-intensive foreign goods. The import of cheaper goods may cause disruption and loss of employment in the developed country. The theory of comparative advantage has suggested, however, that these resources in the developed countries could be diverted to other activities where advantages exist with the result that everyone will be better off. But, it is not as simple as has been suggested because of entrenched and vested interests. This problem will persist as long as developed countries do little to make adjustments or give assistance to help industries within their own countries.

Domestic economic policies in developed countries play a vital role in determining developing country export growth so long as the latter are reliant on the former. The export performance of developing countries is tied directly to the growth and price stability in developed countries. Any downturn in economic activity, a rise in inflation or any economic disruption will rebound more substantially and much sooner on developing countries that are weakest, most vulnerable and most dependent on the developed world. In times of such economic stress, developed countries cannot be blamed for looking after their own. In order to prevent this there is a form of preferential treatment especially for developing countries whose performance in international trade is weak. Other developing countries should pursue policies that provide a measure of self-reliance and to be cautious when cooperating with MNCs in setting up manufacturing plants, transferring technology and joint venture schemes.

Global trade reform

Recent studies by the IMF and the World Bank suggest that trade protection by developed countries costs twice as much as developing countries receive in average annual aid flows. It has been estimated that if the protective barriers were lifted exports from developing countries would rise by about 10 per cent. Other estimates suggest a one-off boost of about 3 per cent in GNP in developing countries. More recent trends point towards the situation where developed countries are now asking for direct and indirect trade reciprocity, for an end to the special and differential treatment accorded to developing countries and to 'graduate' NICs so that these newly 'developed' countries can take their place and play their part in international trade and forfeit the special treatment which they have thus far enjoyed as developing countries. The reality is that increased protectionism is largely the result of greater competition in world markets and that this is further exacerbated by the inability of international agreements to control non-tariff barriers. However, some NICs may be in financial crisis themselves (see Case study G, p. 136).

The use of various types of trade weapons from sanctions to anti-dumping actions, import quotas, the abolition of developing country duty concessions and the removal of most favoured nation trading status to protect primary producers indicates that many countries are jealously guarding their markets. There is a need for global trade reform since the international trading community cannot afford to go on protecting markets into the near future because of increased competition, diminishing returns and market inefficiency.

It has been suggested that the setting up of regional trading blocs will be a step towards global free trade. The argument in favour of the formation of a trading bloc is that it leads to net gains for its members since what was formerly produced domestically may now be imported from lower-cost partners. The other sources of gain include economies of scale in production and increasing competition from larger markets. However, there will be losses if members substitute lower-priced goods from outside the bloc with more expensive goods produced by other members.

The formation of a trade bloc has advantages other than those benefits arising from unilateral trade reform. Developing country trading blocs may benefit primary product exporters through rising exports, but where there are inter-regional rivalries, internal trade may not be easily liberalized. For example, in an Andean Pact, certain markets were allocated to designated producers instead of allowing market forces to determine the allocation of production. These designated producers were neither the most efficient producers nor were tariffs low enough with the result that the overall benefits were restricted.

In trading blocs among developing countries there are usually high tariffs and quotas against non-members and it is likely that the net gains arising from these policy instruments will exceed net losses. However, because trading blocs occupy similar geographical locations, similar goods are produced thus limiting opportunities to exploit differences in skills or endowments. While trade liberalization and multilateral efforts to free up global trade are preferable to the formation of trade blocs, the reality is that trading blocs are but a step towards global trade reform.

In global terms, the three most important trading areas are North America, the Pacific Rim and the European Community/Union (EC or EU). The countries that make up this 'triad' account for about 80 per cent of world GDP and 75 per cent of world imports and exports. US exports to other triad countries accounted for over 81 per cent of total US exports in 1990 (Searing 1992).

The US–Mexico Framework Agreement of 1987 was an initial step taken to integrate the North American market by formalizing all commercial relations between the two countries. The US–Canada Free Trade Agreement (signed in January 1989) created the world's largest trading bloc and will eliminate all tariffs between the two countries over a 10-year period and liberalize investment regulations. These agreements subsequently led to the creation in June 1991 of the North American Free Trade Agreement (NAFTA) designed to eliminate all trade barriers and impediments to investment in the three countries in North America.

Indeed, the former US President Bush's call for an 'Enterprise for the Americas' initiative, since implemented by President Clinton, extends NAFTA down to Central and South America. The key elements of the initiative include working towards a free trade agreement, improved capital flows and investments and programmes to assist indebted nations reduce their debts. This initiative could challenge international efforts at trade liberalization. NAFTA alone is a sizeable trading bloc with a market of about 360 million people and a total economic product of US$6 trillion. This bloc will have both trade diverting effects as well as trade creating effects and will be a substantial challenge to all countries that presently trade with the individual countries of NAFTA. Trade diversion impacts would fall heavily on exports especially primary products as well as manufactured goods such as data processing and telecommunication components. Trade diversion occurs when imports from a preferred trading partner displace imports from a lower-cost non-preferred partner. Trade creation arises when the preferred imports are an addition to imports without regional trade arrangements. That is, they displace domestic production or they arise because lower prices induce greater consumption of a good in the importing country.

The 12 countries of the EU worked towards a single internal market commonly known as EC Project 92, which came into force on 1 January 1992. The single market has removed all barriers to the free movement of goods, services, capital and people among member states. Also because of geographical proximity, the EU has negotiated an economic cooperative agreement with the European Free Trade Association (EFTA) to remove bilateral trade barriers. The EU has also offered Mediterranean and North African countries 'proximity policies' in order to reduce the disadvantages of non-membership. There seem to be greater efforts at resolving internal integration issues than reducing trade barriers to the rest of the world.

In the Pacific Rim, the record of the Association of Southeast Asian Nations (ASEAN) on economic cooperation has been limited. ASEAN has established a programme of gradual intra-ASEAN tariff reduction and introduced schemes such as the ASEAN Industrial Joint Venture (AIJV) scheme which grants tariff relief to ventures in which at least two member states are participants. Intra-ASEAN trade rose from 34 per cent in 1986 to 42 per cent in 1989 and is expected to be about 55 per cent by the year 2000. The proposed ASEAN Free Trade Area (AFTA) is based on membership of ASEAN, that is, the original five countries of Indonesia, Malaysia, the Philippines, Singapore and Thailand that signed the Treaty of Amity and Cooperation in 1965, and Brunei which signed in 1984. AFTA is an increasingly prosperous market of nearly 320 million people, approximately the same size as both the EU or NAFTA. The main push towards free trade is the 'Common Effective Preferential Tariff Scheme' in which tariffs on exports within the region will be reduced to a maximum of 5 per cent at the end of 15 years. Already there is a list of 15 product categories identified for tariff reduction during the first seven years such as vegetable oils, cement, chemicals, pharmaceuticals, fertilizers, plastics, rubber products, leather products, textiles, ceramics and glass, copper and wooden and rattan furniture. Members may submit lists of products for exclusion. To illustrate the relatively small size of ASEAN economies, the six ASEAN countries collectively produce only about 1.4 per cent of world output and 4 per cent of world trade, and, except for Singapore, these countries are at an early stage of economic development.

On a wider regional scale, the Asia-Pacific Economic Cooperation (APEC) was launched in Canberra in 1989 as a forum to integrate US and East Asian economies. Made up of most countries of Asia and the Pacific – Japan, South Korea, Taiwan, Hong Kong, China, the ASEAN states, Papua New Guinea, Australia and New Zealand – APEC includes the fastest growing economies in the world. APEC is seen as the biggest regional grouping in the world accounting for 40 per cent of global trade and nearly 50 per cent of global GNP. This grouping, however, does not imply a full free trade area with minimum internal tariffs. Internationally, it will still rely on the GATT as the world moves towards greater integration in the production of goods, but regionally there will be attempts to remove legal, administrative and infrastructural impediments to trade and investment. With a market of nearly 2 billion people and producing half the world's output it would require all sorts of rules in order to achieve harmonized trade, investment, standards and certification and agreed ways of dispute resolution among member countries. As APEC is built on an economic cooperation model, the appeal of regional trading blocs would be lessened while advancing the regime of trade liberalization and integration.

The above discussion shows that the trend is towards an integration within and across 'triad' areas as evidenced by increasing international mergers, acquisition and direct investment activities. US companies are acquiring European companies as European companies are investing in the US. Japanese investors are buying and investing on both sides of the Atlantic. In these all the nations involved are preserving their own economic and geostrategic interests (see Figure 5.2).

The Japanese in Europe: total investment

US$ million	Year to March 1988	Year to March 1989	Cumulative total 1951–89
UK	2,473	3,956	10,554
Netherlands	829	2,359	5,525
Luxembourg	1,764	657	4,729
West Germany	403	409	2,364
France	330	463	1,764
Switzerland	224	454	1,432
Spain	283	161	1,045
Belgium	70	164	1,027
Ireland	58	42	432
Italy	59	108	370

The Japanese in Europe. Number of manufacturing plants (to January 1989)

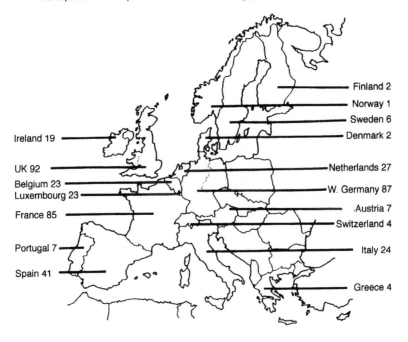

Figure 5.2 Japanese investments in Europe
Source: European Affairs, vol. 3 (Autumn 1989): 98

Case study G

Crisis of the NICs

Walden Bello (1990) outlined in his book *Dragons in Distress: Asia's Miracle Economies in Crisis* some very telling reasons why NICs are in crisis. In 1990 Taiwan recorded its lowest annual rise in GNP since 1982 and South Korea suffered a US$21 billion deficit on its current account after five years of surplus. The crisis of the NIC strategy of export-oriented trade and high-speed growth is said to stem from the intersection of three trends:

- the deterioration of the external trade environment because of rising protectionism;
- the loss of export competitiveness as NICs continue to depend on labour-intensive production even when labour is no longer cheap; and,
- the eruption of long-suppressed environmental, agricultural and political costs exacted by previous rapid growth.

Many of the 'tiger economies' still remain structured on the previous pattern of export-led growth at a time when many markets, especially the US, European and Japanese, are becoming increasingly protected. The markets in eastern Europe and the Commonwealth of Independent States are hardly significant in either the short or medium term given their depressed purchasing power and developing economies. Even without trade barriers or appreciation in value of their currencies, NIC economies have lost their competitive edge as MNCs move elsewhere in search of cheaper labour, and as the product cycle reaches maturity so that goods that were previously profitable and exclusive are now less profitable and are being produced for the mass market. Higher wages have pushed up the cost of living, rural labour reserves have dried up, and labour becoming organized, it has meant, for example, that many South Korean and Taiwanese manufacturers have relocated their operations elsewhere in South East Asia and China. The rapid growth rates in the NIC economies have been won at great cost. Notwithstanding the authoritarianism in South Korea and Taiwan, and the subsequent eruption in 1987 in South

Case study G *(continued)*

Korea, after two decades of tight control, of new democratic policies which have forced responsibility and accountability, environmental degradation itself is threatening productive capacity. In Taiwan, the pollution of rivers by upstream industries has devastated the shrimp, oyster and fish aquaculture industry downstream. Environmental activists in Taiwan have succeeded in delaying the construction of more petrochemical and nuclear power plants – seen as necessities by technocrats to sustain export-oriented growth. Such developments have forced a re-evaluation of benefits of past growth in comparison with the value of personal freedom and quality-of-life issues.

The alternative economic strategy consists of some of the following elements:

- income redistribution in order to expand domestic markets;
- forging a 'New Deal' for agriculture;
- encouraging enterprises that combine profitability with ecological sustainability;
- establishing preferential trading and technological arrangements with as many countries as possible to counteract protectionism;
- enlarging R&D efforts and upgrading technical education;
- devising a targeted export policy which has the potential to provide a firm industrial and technological base for selected manufactured exports; and,
- institutionalizing democratic decision-making in economic policies.

Some of the alternative economic strategies available to NICs are already being used by some countries. But for the majority of developing countries, whether the present regimes have the political will to move in the enlightened direction will depend to a large extent on the degree of resistance and vested interests. This process of renovation may lead to the unravelling of the NIC model that may leave countries other than the Four Dragons in an indeterminate zone between the developed and developing worlds and may indeed cause them to slide back to the Third World in the twenty-first century if development plans are unsuccessful.

The General Agreement on Tariffs and Trade (GATT)

The GATT is the only multilateral institution that lays down agreed rules for international trade. Established in 1948 with 23 signatories, the GATT in 1992 had 117 signatories. The basic aim is to liberalize world trade and contribute to economic growth and development. The GATT is both a code of rules and a negotiating body. In order to liberalize world trade the GATT operates according to a set of principles, the important ones being:

- Most favoured nation (MFN). Trade must be conducted on the basis of non-discrimination as embodied in the MFN clause where no country is to give special trading advantages to another or to discriminate against it. Exceptions to this rule are allowed only under special circumstances.
- National treatment. If a country chooses to give protection to a domestic industry, this protection should be provided through a customs tariff and not through other commercial barriers.
- Transparency. This ensures that traders are not handicapped because of a lack of information on regulations or taxation. Also a stable and predictable basis for trade is provided by the 'binding' of the tariff levels negotiated among the contracting parties. When a country becomes a GATT member, it agrees to fix its tariffs which can only be raised if compensation is negotiated under special terms.
- Consultation, conciliation and dispute settlement are fundamental to the work of the GATT.
- A country may, when its economic or trade circumstances warrant it, seek exemption from certain GATT obligations.
- There is a general prohibition on the use of quantitative restrictions, for example, import quotas, except when a country's balance of payments so warrants.
- The GATT permits the establishment of regional trading arrangements such as the EU or US–Canada Free Trade Agreement provided certain criteria are met. The criteria are meant to ensure that such trade arrangements do not raise barriers to world trade (Searing 1992: 101–3).

The way in which these principles are set in place is through a series of multilateral trade negotiations. Since 1947 eight rounds of trade negotiations have taken place (see Figure 5.3).

While the early rounds of negotiations achieved reductions in tariffs

✿ 1947 **GENEVA ROUND**
Participating countries: 23
Result Summary: Agreement on 20 schedules covering 45,000 tariff concessions. Manufacturing tariff in industrialized countries started dropping from average of 40 per cent to 25 per cent over decade.

✿ 1948 **ANNECY ROUND**
Participating countries: 13
Result Summary: Additional 5,000 tariff concessions exchanged to continue the tariff cutting momentum.

✿ 1950 **TORQUAY ROUND**
Participating countries: 38
Result Summary: 8,700 tariff concessions exchanged.

✿ 1956 **GENEVA ROUND**
Participating countries: 26
Result Summary: About US$2.5 billion worth of tariff reductions.

✿ 1960–1 **DILLON ROUND**
Participating countries: 26
Result Summary: 4,400 tariff concessions covering US$4.9 billion worth of trade.

✿ 1964–7 **KENNEDY ROUND**
Participating countries: 62
Result Summary: Introduced across-the-board rather than product-by-product approach to cutting tariffs. Concessions covering US$40 billion worth of trade. Anti-dumping rules introduced. By end of decade, industrial tariffs averaging about 17 per cent.

✿ 1973–9 **TOKYO ROUND**
Participating countries: 99
Result Summary: Tariffs reductions covering US$300 billion of trade. Preferential treatment for developing countries. Codes agreed on subsidies, countervailing measures and other non-tariff barriers. Led to progressive lowering of industrial tariffs to current level of 4.7 per cent.

✿ 1986-93 **URUGUAY ROUND**
Participating countries: 116
Result Summary: If successful, will bring range of important trading areas: agriculture, services, intellectual property rights and investment -- GATT will become the Multilateral Trade Organization with re-vamped rules and dispute-setting machinery.

Figure 5.3 The GATT and trade negotiation rounds

on thousands of goods the Tokyo Round went substantially beyond traditional negotiations that eliminated tariffs and quotas. The Tokyo Round covered eight areas of contention:

- An improved legal framework for the conduct of world trade including the establishment of the generalized system of preferences (GSP) under which developed countries gave preferential tariff and non-tariff treatment to developing countries.
- Non-tariff measures such as subsidies and countervailing duties.
- Technical barriers to trade.
- Government procurement.
- Customs valuation.
- Import licensing procedures.
- Revision of the GATT anti-dumping code.
- The bovine meat, dairy products, tropical products and civil aircraft industries.

As a result of the Tokyo Round the US GSP provided duty-free tariff preferences for developing countries. It recognized that developing countries needed preferential treatment to compete effectively with industrial countries in the US market. Under this scheme, trade rather than aid is seen as a more effective and cost-efficient means for promoting broad based economic growth and export-led industrialization. Currently the US offers duty-free preferential treatment on 4,100 tariff categories to 128 beneficiary countries and territories. The US Trade and Tariff Act of 1984 requires the president to evaluate a country's trade-related practices before granting or maintaining GSP benefits. Some of the criteria used to evaluate trade practices include workers' rights, intellectual property rights protection, the extent barriers to trade in services are eliminated and the reduction of trade-distorting investment practices. Nicaragua and Romania were removed from the list, while Chile, Paraguay, the Central African Republic and Myanmar were suspended due to violation of workers' rights and the benefits to Thailand were reduced for failing to provide adequate protection to intellectual property rights. More significantly, in 1989, South Korea, Taiwan, Hong Kong and Singapore – the NICs – were 'graduated' from the programme since these economies were deemed to be able to compete effectively in the US market without trade preferences.

Non-tariff barriers include direct trade measures such as variable levies, quantitative restrictions, import controls, countervailing duties and export subsidies. Such barriers violate the GATT principle that protection should only be by means of customs tariffs. Technical barriers include regulation and standards for reasons of safety, health and environmental protection. Classic examples are Japan's regulation requiring *all* aluminium baseball bats to be split to check their cores, thus rendering the bats almost unsaleable, or the US strict emission controls for imported cars. Thus government procurement policies may appear open to competition but are often exclusionary and unfair in practice. The US is seeking greater transparency in the rules and regulations for bidding on government contracts especially in developing countries so that the process and criteria for the award of contracts is fully disclosed and fair. These are all considered new forms of protectionism – the so-called non-transparent measures.

The Uruguay Round, by far the most ambitious and comprehensive of all the GATT Rounds, will establish a new order in world trade through the Multilateral Trade Organization (MTO). Developing countries sought to include temperate and tropical agriculture, textiles and clothing within the GATT, and to improve rules affecting safeguards such as anti-dumping and countervailing duties, dispute settlement and the functioning of the GATT rules. Industrial countries introduced trade in services, trade-related investment measures and intellectual property rights and protection.

While the dynamic NICs were interested in market access for their exports, the largest group of developing countries represented by the small, vulnerable economies with serious debt problems, and fragile, distorted economic structures, were eager to be included in world trade.

International trade in services accounts for an increasing percentage of world trade. Such services include banking, insurance, transport, communications, computer services, tourism, construction and education services. However, because the GATT was originally intended to cover trade in goods, trade in services has fallen outside its jurisdiction. Hence, special negotiations for services have had to be established. The protection of intellectual property rights including patents, book copyright, trademarks and computer software has been inadequate and proponents have sought stronger rules and more effective measures.

Trade-related investment measures refer to government regulations and policies on investors that distort and restrict investment flows and patterns of trade. For example, the linking of local employees and local content rules have been used widely in South East Asia, the Caribbean and South American countries. Agreement has been reached in the target amounts for tariff reduction and a strengthening of the GATT rules and procedures which will improve the effectiveness of the GATT dispute settlement process, trade-related investment measures and market access in natural resource based products.

However, the inability to agree on substantive reform of agricultural policies led to a breakdown of negotiations in mid-December 1990 in Brussels. The US, EU, Japan and South Korea were unable to agree on agricultural policies such as reform of world farm trade, liberalization of services and market access. For instance, reductions in export subsidies, internal support and market access in the EU were contentious issues that were only resolved after prolonged negotiations.

The threat of regional trading blocs has underlined the importance of a rules-based multilateral trading system. While previously developing countries were satisfied with a 'separate and different' treatment, trends suggest a jettisoning of such ideas. Some Latin American countries and India, which have anti-trade biases in their development policies, are keen to retain their separate and different status while others are negotiating new trade agreements. The poorest developing countries like those in Sahel Africa have no interest in the GATT negotiations and rely on aid and preferential access for exports. On the other hand, East Asian countries have the strongest interests in the trade system since they depend on export growth for their development and a liberal multilateral trade system suits their interests best. While the US is keen to pursue its free trade policies without imposing new barriers on to its trade partners, the EU in its pursuit of a single market, will attempt to reduce its support for agriculture by re-examining its Common Agricultural Policy (CAP).

The outcome of the concluded Uruguay Round will lead to better market access for both industrial and developing countries. There will be lower tariffs world-wide, significant reductions in agricultural and industrial subsidies, an extension of multilateral arrangements to services, trade-related investment rules and intellectual property rights. The only question is the size and speed of restrictions in export subsidies, changes to domestic price supports and lowering of import barriers. Progress has already been made in some sectors like textiles and clothing, services, tariff cuts, intellectual property rights and dispute settlement processes. The successful Uruguay Round is therefore crucial in ensuring a world trading system that will lead to strong global economic growth.

A General Agreement on Trade in Services (GATS)

Expanding global trade and investment flows has meant that a separately negotiated General Agreement on Trade in Services (GATS) has had to be discussed apart from the Uruguay GATT negotiations. The intention is to create a GATT-styled regime for services. In 1993 the world services trade accounted for US$850 billion, about 20 per cent of total world trade. Services are a key 'value-adder' in many economies and there is a trend towards the internationalization of services. Services are also the fastest growing single component of world trade, increasing by 66 per cent in the period between 1985 and 1990. The services

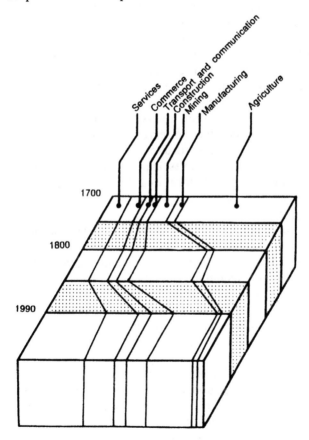

Figure 5.4 The labour shift into services

sector includes financial services, telecommunications, aviation and consultancy services among others (see Figure 5.4).

There are three components to the negotiations in services. The first is the need to establish a GATS framework that will establish international rules that govern trade in services similar to existing GATT rules for trade in agricultural and manufactured goods. The second component is to complete a package of agreements on market access in the services sector between all GATT signatory countries. The final component is that countries make guarantees on market access and national treatment for specific sectors.

Within the proposed framework there are two principal rules, namely the most favoured nation clause and national treatment. As in the GATT, the most favoured nation clause states that no country can give favoured access to the services market within its own territory to another country without extending the same condition to others from all GATT countries. This clause ensures non-discrimination and that all firms can compete on equal terms in the services export markets so that actions that affect the trade of one country should apply comparably to all GATT participants. The second rule refers to national treatment which stipulates that governments extend to service providers treatment which is no less favourable than that accorded to its own domestic services and service providers in so far as taxes and regulations are concerned. However, unlike the first rule, national treatment is an obligation which only applies to those service sectors where countries choose to make market access guarantees.

Other minor rules that need clarification and agreement include the requirement for transparency, the provision for the recognition of qualifications and disciplinary codes on service monopolies. The intention of the transparency requirement is to ensure that governments are open in all their international dealings. How governments intervene in the market must be clear to all. This means that comprehensive and reliable information on laws and regulations are freely available to all, that government policies and guidelines are freely available and that government policies and guidelines applying to services in targeted export markets are similar for both domestic and international firms.

The GATS should also bind signatory nations to either eliminate or freeze measures that limit foreign access to particular service sectors. The intended effect of all these measures is to ensure that access to the services market world-wide is open and non-discriminatory. In bilateral negotiations the aim will be to identify and remove barriers to trade in services. Such barriers include bans on new entrants to markets, the limitation on the types of business activities foreign firms may carry out, the more severe financial conditions that are placed on foreign firms planning to establish themselves in domestic markets and the limitations placed on the temporary entry of service providers in the domestic market.

As will be noted, there are vast opportunities for trade in services given the growth of the world economy and the relative affluence of industrial countries. In the case of Japan, tourism is a strong performer given a relatively affluent ageing Japanese population combined with new attitudes to work and leisure. Other East Asian countries exhibit growing opportunities in different sectors. Hong Kong is a fertile market for high value-added services, notably engineering services, infrastructural development, building and construction and the environment such as waste management practices. On the other hand, in Taiwan the growing service sectors are tourism, education services and environmental management, whereas in southern China the growth service sectors are telecommunications and education. In general, for most of South East Asia, the stronger service sectors seem to be in telecommunications, education and training, transport, building and construction and in health and medical services. Different countries within the region will require different types of services, a reflection of the variable stages of development and needs.

International trade and ecologically sustainable development (ESD)

Economic development and environmental conservation need not be mutually exclusive (World Bank 1991). There are a wide range of environmental actions that have high returns. The GATT does not mention the environment in its terms of reference. While this may be a major gap, too much may be made of it – for example, the Antarctic Treaty scarcely mentions the environment but a great deal of effort is made to protect the Antarctic environment. The environmental question has been given scant attention and priority in the GATT: rather the emphasis has been on the trade impacts of pollution as against broader environmental issues. However, more recently, the trade effects on the environment have been attracting the increasing attention of the international community. Such attention is polarized between the traders, on the one hand, being concerned that environmental policies might affect trade competitiveness while, on the other, the environmentalists fear that trade policies may encourage further environmental damage. So the challenge for governments is to identify the links between international trade and environmental policies and to find a way to integrate them.

Included in this quest is the identification of where economic and environmental objectives are mutually compatible and supportive of each other as well as where there may be conflicts between the two. The environmental question in both national and international debates has important implications for trade. For example, international trade reflects the international market and since this market may not reflect the social costs of environmental damage in the production or consumption of goods traded, trade and environmental objectives may be in conflict.

In the 1990s and beyond and especially after the UN Conference on Environment and Development (UNCED) in Rio in June 1992 – the Earth Summit – the catchcry is for an ecologically sustainable development (ESD) of countries, of economies and of resources. This implies that development meets the needs of the present without compromising the ability of future generations to meet their own needs (World Commission on Environment and Development (Brundtland Report) 1987: 43). These ideas have found legislative effect in some countries. For example, New Zealand's Resource Management Act 1991 defined 'sustainable management' in s.5 as 'managing the use, development and protection of natural and physical resources in a way, or at a rate, which enables people and communities to provide for their social, economic and cultural well-being and for their health and safety'. This statutory response contains three fundamental axioms in which the use, development and protection of resources can take place, namely that:

- the potential resources to meet the needs of future generations are sustained;
- the life-supporting capacity of air, water, soil and ecosystems is safeguarded; and,
- any adverse effects of activities on the environment are avoided, remedied or mitigated.

However, the nexus between ESD and international trade is not as straightforward as suggested by the ideal of sustainability. The extent and nature of trade is important for evaluating a country's resource use. For example, Japan may switch from importing bauxite and coal from Australia to importing aluminium. How will this change be accounted for in energy use, in depletion of resources and in adding to the pollution of the environment is a difficult conceptual and statistical problem. But the larger issue is that while one country may achieve sustainability along one path of development, it may be at a cost of non-sustainability in another sector. A further example is the hardwood tropical timber trade which suggests that there may be a drawing down of natural resources of developing countries exporting these timbers even though claims as to its sustainability have been made. Indeed, a more severe indictment of this trade is that developing countries may in effect be subsidizing exports of tropical timber to developed countries, either because of competitive pricing in order to secure markets or because it is nearly impossible to sustain such harvests in the short to medium term.

Thus, for international trade to be a significant contributor to ESD there needs to be international action on environmental public goods, national actions on domestic industrial processes and on final products. Such an integrated approach may ensure better decision-making in trade and environmental matters. National ESD without trade would be feasible, but sometimes very costly. In other cases, ESD without trade would be hardly feasible, for example, in an entrepôt trading nation like Singapore, where its economy depends almost entirely on trade. The dilemma therefore is that while international trade improves the efficiency with which resources are used, the trade impact may be disproportionate to the environmental gain. On the other hand, if environmental objectives were to be achieved, the effect on trade may be very severe and costly, and international trade distorted. The relationship between trade and the environment is thus dependent and conditional, given that a change in one will impact on the other.

The Brundtland Report (1987) has suggested that protectionism increases the likelihood of environmental degradation. In industrial countries protectionism stifles export growth and prevents diversification in developing countries away from traditional and often environmentally damaging exports. For example, the shift away from traditional agriculture to a European-styled agriculture encourages the excessive use of chemical fertilizers and pesticides.

The experience of trade liberalization has not been good for the environment. For example, the harmonization of standards in the EU means that the Danish law demanding that all bottles be returnable or recyclable be abandoned. This follows a European Court decision which held that while the Danish approach was good for the environment, nevertheless it kept other European bottle manufacturers out of

Denmark. This conflict between free trade and environmental priorities has also been encapsulated in the attempts by the US to ban the trade in canned tuna because dolphins were dying in tuna nets. The US Marine Mammals Protection Act 1972 was amended to set quotas and permissible numbers of dolphin deaths for tuna boats. Mexican boats were allowed to kill 25 per cent more dolphins than US boats. Even so, Mexico complained to the GATT in 1991 which held that the ban was an infringement to free trade and that Mexican tuna exports should be allowed to resume unfettered. The outcome of this decision now means that countries are unable to use trade as an environmental lever and implicitly accords trade a higher priority than the environment. It also means that the GATT judgement encourages the continuation of environmentally unfriendly trade and the continued degradation of the environment.

It has been argued that free trade will lead to the relocation of polluting industries to poorer countries, renewed deforestation in favour of agriculture and the loss of policy standards that governments use to maintain environmental standards. However, the proponents of free trade argue that these fears may be misplaced. Import barriers and government subsidies both severely distort international trade. This is because these measures often lead to massive logging and deforestation, the over-use of chemicals and pesticides and to poverty in general. Treaties such as the GATT lower barriers and subsidies and can actually reduce damage to the environment more effectively than other agreements targeted specifically at the environment, for example, agriculture. The GATT, by reducing or eliminating trade distortions, increases the efficiency with which the resources are used and increases economic welfare. These achievements, however, require the elimination of unsustainable uses of the environment as well as government intervention with economic or other policy sanctions.

Case study H

Trade and the environment

In an age of environmental enlightenment, pressing questions regarding the link between trade and the environment have often been asked. Some of these questions include:

Case study H *(continued)*

- What are the effects of trade liberalization on the environment?
- How can trade policies be used to influence environmental standards?
- What is the role of trade policies in enforcing international environmental agreements?

The effect of trade liberalization on the environment is one of fear and apprehension since there is a belief that in striving for greater and greater shares of international trade, environmental goals are either pushed into the background or ignored totally. For example, with the easing of trading restrictions in North America through NAFTA, air and water quality in both Canada and Mexico will be seriously undermined through trans-border pollution; allowing cassava into the EU may have a deleterious effect on soil erosion in Thailand and other cassava-producing countries. On the other hand, it may be argued that trade liberalization may introduce greater efficiency and higher productivity through the use of cleaner technology and less polluting industries. However, trade policies are blunt instruments for environmental management because resource use is influenced indirectly. A more potent mechanism is to use market forces to provide a leverage for environmental management because if the environment was 'priced' appropriately, then industries would take more heed of environmental degradation such as deforestation, soil erosion and air and water pollution.

It has been argued that developed countries strive to achieve high environmental standards in order to be model international good citizens for others to emulate. A corollary to this view is that such countries use these standards to protect domestic industries from foreign competition. The argument, however, is a spurious one because the high standards are designed to prevent environmental damage generally. Consumption patterns in developed countries are such that a high standard is required in order to maintain the expected high quality of life. Moreover, in these countries the priorities, the capacity to assimilate pollutants and environmental degradation are already in place to counter any negative effects from increased consumption. On the other hand, the view that high environmental standards in developed countries push polluting industries offshore to 'pollution havens' in

---

Case study H *(continued)*

developing countries cannot be supported. Foreign investment flows have not relocated to countries where there are lax environmental standards. Indeed, there is anecdotal evidence from Chile which suggests that MNCs using the same technology as in developed countries have reduced pollution levels.

In the environmental area, trade sanctions have sometimes been used as a weapon to encourage environmental awareness among countries. The trade in ivory, the export of rain-forest hardwoods and the use of chemicals that deplete the ozone layer are some examples where trade sanctions have been exercised. International agreements cannot be enforced if the offending country has not been a signatory or a party to that agreement. Thus, it seems that the use of trade instruments could be justified in some cases. Some well-known international agreements include CITES (1973), an agreement on the international trade in endangered species of wild fauna and flora which has placed an embargo on the ivory trade and rhinoceros horns, the Montreal Protocol (1987) to phase out ozone-depleting chemicals and the Basel Convention (1992) to control the transboundary movement and disposal of hazardous wastes.

The ban on the trade in ivory is aimed at protecting the African elephant, but this involves complex difficulties and trade-offs. A ban on the trade may drive prices up and induce more poaching. In countries like Botswana, South Africa and Zimbabwe big game hunting and safaris help enrich the local population and bring revenue to government agencies. Such revenues can be used for managing the elephant herds and to finance law enforcement. This argument, however, masks a more difficult conservation dilemma, that is, who are we saving the elephants from and what are its benefits? The answer to the first is to save the elephant from mankind and to ensure the preservation of the species. This altruistic purpose involves a degree of self-sacrifice because the ultimate beneficiary is the world community. The personal loss to those who depend on the trade based on the elephant would have to be made up by other means such as international assistance and passive tourism where the only shooting that occurs is through the lens of a camera. Thus, it can be seen that trade policies can be used to implement international environmental agreements.

Conclusion

Developing countries have to contend with a global financial and trading system which is in a state of flux – a system that is the most obvious and familiar aspect of the economic climate. Of the world's 4 billion or so people, nearly 80 per cent live in developing countries. However, the share of global output from the developing world is less than 20 per cent and of world trade even less at 17 per cent. Thus, as a group, developing economies are still far from being fully integrated into the global economy. There are also other uncertainties which developing countries have to confront, including security, political uncertainty, technological advance, the energy outlook and environmental damage. The challenges are therefore daunting. In 1992 Paul Ekins suggested that a new world order may develop in one of three ways.

1 The Neo-Liberal Order. This world is made up of a global market where an increasing number of people and resources are engaged in exchange relationships. While not a free market both property owners and consumers will be better off under this regime. The main protagonists, of course, will be institutions such as the World Bank, IMF and the GATT.
2 Social Democratic New World Order. Such a state is based on the New International Economic Order (NIEO) where markets are accepted and nation states valued within social and democratic principles. Here the UN and other government bodies such as the EU are an integral part of the order.

Both world orders above are characterized by being western-oriented with the rest of the world entertaining a vision of becoming like the US. Human progress is defined in economic terms such as GNP per capita and planning is essentially a 'top-down' affair which is managed by owners of MNCs and international bureaucracies.

3 Grass-Roots New World Order. In this mode, there is a 'bottom-up' ethic and philosophy initiated by local groups. Such an order is aimed at combating four holocausts – war and militarization, human oppression, economic destitution and environmental degradation. Such a new world order works hand in glove with Peace, Green Development and the Human Rights movements. The main aims are to increase creativity and cultural diversity, to promote holistic development and a world order that is committed to social development, non-violence and the environment.

Whichever model succeeds and dominates, there appears to be no magic cure for economic underdevelopment. There may be more than one way to succeed and an evaluation of success needs to be made according to the various dimensions of development, not merely income growth. The interaction between domestic markets and the global economy are crucial for a new world order to come about. Effective domestic markets have a tendency to attract foreign investments which in turn boost productivity. International trade links also allow countries to make use of comparative advantages thus helping domestic economies make more efficient use of their resources.

The task therefore is to redress the global imbalance where developed countries having only one-fifth of the world's population produce four-fifths of world output, and account for more than four-fifths of world trade. All exports of capital and technology come from the developed world. However, in the pursuit of this goal there is a need to avoid the geopolitician's tripolar world, if only because three is an awkward number which may encourage two to gang up on the third, or one to play the other two off against each other. In the international scene of trade blocs, the triad of NAFTA, EU and the Pacific Rim countries looms large, very much like George Orwell's *1984* where the super-states of Eurasia, Eastasia and Oceania are engaged in global rivalry. In the time it takes to read this paragraph, about 100 children will be born, six in industrial countries and 94 in developing countries. This then is the global challenge. Despite uncertainties, more countries are adopting market-friendly approaches with strong international cooperation to enhance opportunities for global trade and development.

Key ideas

1 Free trade is best when it has the effect of maximizing world economic output. However, before developing countries may fully subscribe to this ideal there may be a need to practise a limited amount of protection of their production and manufactures.
2 Governments protect their markets using various devices for different reasons, including industrial promotion, diversification of exports, the creation of employment and development-oriented goals.
3 As a result of greater competition in international trade, there is an increase in protectionism. International agreements have been unable to control protective non-tariff barriers. There is an immediate need for global trade reform.

4 To achieve free trade there is a need to go through the phase where trading takes place through regional trading blocs. These blocs are a step towards global free trade.
5 Global trade reform through the auspices of the GATT is seen by many as the next logical step in international trade.
6 Economic development and environmental conservation need not be mutually exclusive. International trade and ecologically sustainable development can indeed coexist but the relationship is a dependent and conditional one.

Review questions, references and further reading

Items for further reading are indicated with an asterisk.

Chapter 1

Review questions

1 Discuss what is trade and how it may impact on the various facets of economic and social activity.
2 Trade theory suggests that trade occurs only because of comparative advantage. Examine this notion through time.
3 Trade has brought about dependency while multinational corporations have spawned global production and a new international division of labour. List some advantages and disadvantages of these to developing countries.
4 Foreign aid is both a sufficient and necessary condition for economic development. Discuss.

References and further reading

*Alvstam, C. (1993) 'The impact of foreign direct investment on the geographical pattern of foreign trade flows in Pacific Asia with special reference to

Taiwan', in C. Dixon and D. Drakakis-Smith (eds) *Economic and Social Development in Pacific Asia*, London: Routledge, pp. 63–84.

*Cho, G. and Williams, S.W. (1990) 'Trade, aid and regional integration', in D. Dwyer (ed.) *South East Asian Development: Geographical Perspectives*, London: Longman, pp. 225–55.

El-Agraa, A.M. (1983) *The Theory of International Trade*, London: Croom Helm.

Krugman, P. (1991) *Geography and Trade*, Leuven: Leuven University Press.

Riedel, J. (1988) 'Trade as an engine of growth: theory and evidence', in D. Greenaway (ed.) *Economic Development and International Trade*, Basingstoke: Macmillan, pp. 25–54.

Chapter 2

Review questions

1 Discuss what is meant by 'development' and 'growth' of developing countries and suggest how trade may assist in this process.

2 How may one analyse the structure of trade? Using one method of analysis describe the broad patterns of world trade either for a regional group of countries or for a single country.

3 What trade policies will be useful for newly industrializing countries as opposed to mere developing countries of the Third World?

4 Developing countries cannot rely upon expanding foreign trade as a means of achieving development. Elaborate with specific examples.

References and further reading

Garnaut, R. (1989) *Australia and the Northeast Asia Ascendancy*, Canberra: Australian Government Publishing Service.

Gilpin, R. (1987) *The Political Economy of International Relations*, Princeton, NJ: Princeton University Press.

IMF (1990) *Direction of Trade Statistics Yearbook*, New York: IMF.

Independent Commission on International Development Issues, Brandt Commission (1980) *North–South: A Program for Survival*, Cambridge, MA: MIT Press.

*Johnston, R.J. (1967) *The World Trade System. Some Enquiries into its Spatial Structure*, London: G. Bell & Sons.

*Todaro, M. (1981) *Economic Development in the Third World*, London: Longman.

World Bank (1991) *World Development Report, 1991. The Challenge of Development*, New York: Oxford University Press.

—— (1992) *World Development Report, 1992. Development and the Environment*, New York: Oxford University Press.

Chapter 3

Review questions

1 How important is foreign aid to the economies of the Third World in relation to their other sources of foreign exchange receipts? Explain the various forms of development assistance and distinguish between bilateral and multilateral assistance. Which do you think is more desirable and why?

2 What is meant by 'tied aid'? Both capitalist and socialist nations have increasingly shifted from grants to loans and from untied to tied loans and grants during the past decade. What are the major disadvantages of tied aid especially when this aid comes in the form of interest-bearing loans?

3 Under what conditions and terms do you think LDCs should seek and accept foreign aid in the future? If aid cannot be obtained on such terms, do you think LDCs should accept whatever they can get? Explain your answer.

4 Aid can be effective for economic development. However, some countries cannot 'afford' to accept aid. Critically evaluate.

5 Aid can be anti-developmental. Elaborate with examples.

References and further reading

Corbridge, S. (1992) *Debt and Development*, Oxford: Blackwell.
Hayter, T. (1971) *Aid as Imperialism*, Harmondsworth: Penguin.
*Hayter, T. and Watson, C. (1985) *Aid: Rhetoric and Reality*, London: Pluto.
*Kreuger, A.O, Michalopoulous, C. and Ruttan, V. (1989) *Aid and Development*, Baltimore, MD: Johns Hopkins University Press.
*Remenyi, J.V. (1991) *Where Credit is Due*, London: ITC Publications.
Singer, H. and Ansari, J. (1982) *Rich and Poor Countries*, London: George Allen & Unwin.
World Bank (1990) *World Debt Tables 1990–91. External Debt of Developing Countries*, Washington, DC: The World Bank.
———— (1991) *World Development Report, 1991. The Challenge of Development*, New York: Oxford University Press.
———— (1992) *World Development Report, 1992. Development and the Environment*, New York: Oxford University Press.

Chapter 4

Review questions

1 Countertrade puts off the need for many LDCs to correct the basic causes of poor trade performance. Indeed countertrade is a backward bartering system. Discuss.

2 Countertrade is an inefficient form of trade and may be contrary to the principles embodied in the GATT. Elaborate with examples.
3 The import of technology to achieve economic development and growth may be counterproductive. Analyse with examples from different countries.
4 Absorptive capacity is required before the transfer of technology can be made effective. Discuss.

References and further reading

*Barrera, E. and Williams, F. (1990) 'Mexico and the United States. The Maquiladora Industries', in F. Williams and D. V. Gibson (eds) *Technology Transfer: A Communications Perspective*, Newbury Park, CA: Sage Publications.

*Bello, W. (1990) *Dragons in Distress: Asia's Miracle Economies in Crisis*, San Francisco, CA: Food First.

Carr, M. (ed.) (1985) *The AT Reader. Theory and Practice in Appropriate Technology*, New York: Intermediate Technology Development Corporation of America.

*Chudnovsky, P. (1990) *North–South Technology Transfer Revisited. Research Issues for the 1990s*, Ottawa: International Development Research Centre.

Schaffer, M. (1989) *Winning the Countertrade War: New Export Strategies for America*, Somerset, NJ: John Wiley.

Schumacher, E.F. (1973) *Small is Beautiful: a Study of Economics as if People Mattered*, London: Blond and Briggs.

Vernon, R. (1973) *Multinational Enterprises*, Paris: Colmann-Crug.

World Bank (1993) *The East Asian Miracle: Economic Growth and Public Policy*, New York: Oxford University Press.

Chapter 5

Review questions

1 Policy interventions to promote free trade will be more beneficial than are attempts to protect markets. Elaborate with examples.
2 The growth of regionalism and trade blocs are a symptom that international trade is in need of reform. Do you think that this is a fair assessment?
3 The GATT has been a useful mechanism that controls world trade. Examine its guiding principles and demonstrate how these may or may not operate effectively.
4 Trade and ecologically sustainable development are not necessarily incompatible. For growth and economic development these elements need to coexist in a balanced manner. Give your views for either case.

References and further reading

Bello, W. (1990) *Dragons in Distress: Asia's Miracle Economies in Crisis*, San Francisco, CA: Food First.
Bello, W. and Broad, R. (1992) 'The crisis of the NICs: fundamental not transitional', *Australian Development Studies Network Briefing Paper No. 24* (January), Canberra: Australian National University.
Edwards, C. (1985) *The Fragmented World: Competing Perspectives on Trade, Money and Crisis*, London: Methuen.
Ekins, P. (1992) 'A new world order – for whom?, *Development Bulletin*, vol. 23, p. 35.
Harris, S. (1991) 'International trade, ecologically sustainable development and the GATT', Department of International Relations Working Paper, Research School of Pacific Studies, Canberra: Australian National University.
Orwell, G. (1949) *1984*, London: Secker and Warburg.
Searing, J.E. (1992) *The Ernst & Young Resource Guide to Global Markets, 1992* (2nd edn), New York: John Wiley.
Snape, R., Adams, J. and Morgan, D. (1993) *Regional Trade Agreements. Implications and Options for Australia*, Canberra: Australian Government Publishing Service.
World Bank (1991) *World Development Report 1991. The Challenge of Development*, New York: Oxford University Press.
*World Commission on Environment and Development (1987) *Our Common Future. The Brundtland Report*, Oxford: Oxford University Press.

Index

Milton Keynes UK
Ingram Content Group UK Ltd.
UKHW022359061024
449327UK00031B/2576